Enhanced Coagulation Impacts on Water Treatment Plant Infrastructure

The mission of the Awwa Research Foundation (AwwaRF) is to advance the science of water to improve the quality of life. Funded primarily through annual subscription payments from over 1,000 utilities, consulting firms, and manufacturers in North America and abroad, AwwaRF sponsors research on all aspects of drinking water, including supply and resources, treatment, monitoring and analysis, distribution, management, and health effects.

From its headquarters in Denver, Colorado, the AwwaRF staff directs and supports the efforts of over 700 volunteers, who are the heart of the research program. These volunteers, serving on various boards and committees, use their expertise to select and monitor research studies to benefit the entire drinking water community.

Research findings are disseminated through a number of technology transfer activities, including research reports, conferences, videotape summaries, and periodicals.

Enhanced Coagulation Impacts on Water Treatment Plant Infrastructure

Prepared by:
Marc Edwards and **Paolo Scardina**
Department of Civil Engineering, 407 Durham Hall
Virginia Tech, Blacksburg, VA 24061-0246
and
Laurie S. McNeill
Utah State University, 8200 Old Main Hill, Logan, UT 84322-8200

Sponsored by:
Awwa Research Foundation
6666 West Quincy Avenue, Denver, CO 80235-3098

Published by:

Awwa Research Foundation
Advancing the Science of Water®

IWA Publishing

DISCLAIMER

This study was funded by the Awwa Research Foundation (AwwaRF). AwwaRF assumes no responsibility for the content of the research study reported in this publication or for the opinions or statements of fact expressed in the report. The mention of trade names for commercial products does not represent or imply the approval or endorsement of AwwaRF. This report is presented solely for informational purposes.

Published by IWA Publishing, Alliance House, 12 Caxton Street, London SW1H 0QS, UK

Telephone: +44 (0) 20 7654 5500; Fax: +44 (0) 20 7654 5555; Email: publications@iwap.co.uk;
Web: **www.iwapublishing.com**

AwwaRF report number 90997F. Originally published by AwwaRF for its subscribers in 2003.
IWA Publishing version published 2004.

ISBN 1 84339 8931

CONTENTS

TABLES

FIGURES

FOREWORD

The Awwa Research Foundation is a nonprofit corporation that is dedicated to the implementation of a research effort to help utilities respond to regulatory requirements and traditional high-priority concerns of the industry. The research agenda is developed through a process of consultation with subscribers and drinking water working professionals. Under the umbrella of a Strategic Research Plan, the Research Advisory Council prioritizes the suggested projects based upon current and future needs, applicability, and past work; the recommendations are forwarded to the Board of Trustees for final selection. The foundation also sponsors research projects through the unsolicited proposal process; the Collaborative Research, Research Applications, and Tailored Collaboration programs; and various joint research efforts with organizations such as the U.S. Environmental Protection Agency, the U.S. Bureau of Reclamation, and the Association of California Water Agencies.

This publication is a result of one of these sponsored studies, and it is hoped that its findings will be applied in communities throughout the world. The following report serves not only as a means of communicating the results of the water industry's centralized research program, but also as a tool to enlist the further support of the nonmember utilities and individuals.

Projects are managed closely from their inception to the final report by the foundation's staff and large cadre of volunteers who willingly contribute their time and expertise. The foundation serves a planning and management function and awards contracts to other institutions such as water utilities, universities, and engineering firms. The funding for this research effort comes primarily from the Subscription Program, through which water utilities subscribe to the research program and make an annual payment proportionate to the volume of water they deliver and consultants and manufacturers subscribe based on their annual billings. The program offers a cost-effective and fair method for funding research in the public interest.

A broad spectrum of water supply issues is addressed by the foundation's research agenda: resources, treatment and operations, distribution and storage, water quality and analysis, toxicology, economics, and management. The ultimate purpose of the coordinated effort is to assist water suppliers to provide the highest possible quality of water economically and reliably. The true benefits are realized when the results are implemented at the utility level. The foundation's trustees are pleased to offer this publication as a contribution toward that end.

Edmund G. Archuleta. P.E.
Chair, Board of Trustees
Awwa Research Foundation

James F. Manwaring, P.E.
Executive Director
Awwa Research Foundation

ACKNOWLEDGMENTS

This work was supported by a grant from the Awwa Research Foundation (AwwaRF). We appreciate the extensive insights and constructive criticism provided by our project advisory committee members: Timothy D. Barr (Washington Suburban Sanitary Commission); Dominique Gatel (Water Treatment Research Manager); Malcolm Taylor (Buchart Horn, Inc.); and Christian Volk (Indiana-American Water Co.). Special thanks to the hundreds of utility personnel and experts from around the world that participated in the surveys and case studies presented in this project-- this product directly results from their experiences, cooperation and efforts.

EXECUTIVE SUMMARY

"America's water and wastewater utilities face an estimated funding gap of $23 billion per year between current investments in infrastructure and the investments that will be needed...water systems account for $11 billion per year." The Water Infrastructure Network (AWWA, 2001)

Production of safe drinking water is the steadfast goal of the water industry, but achieving that goal at reasonable cost is increasingly a challenge. Over the next 20 years, utility spending on infrastructure will be at least twice that spent on water treatment. Occasionally, new USEPA regulations require changes to water treatment that can even accelerate infrastructure degradation. One recent regulation of particular concern in this regard is the Stage 1 requirements of the EPA Disinfection By-Products Rule, which can require improved removal of TOC from water supplies. One of the most common means of improving TOC removal is to enhance existing coagulation treatment processes by reducing coagulation pH or using higher coagulant doses. There is substantial concern that the lowered coagulation pHs and higher coagulant doses will significantly accelerated degradation of infrastructure of drinking water treatment plants.

The Awwa Research Foundation (AwwaRF) Project "Effects of Enhanced Coagulation on Degradation of Drinking Water Treatment Plant Infrastructure" was designed to produce several important products for utilities. Since very little research has been conducted into infrastructure degradation from coagulation, the first step was to compile insights from other industries in a comprehensive literature review. Thereafter, attempts were made to gain expertise from direct experiences of the Metropolitan Water District of Southern California, international experts from France, Australia and the United Kingdom, and a survey of coagulation utilities in the United States. Some direct experimentation was also conducted to answer a few key questions, including potential inhibitor compatibility with enhanced coagulation and a direct head to head comparison of alum, ferric and PACl coagulants in degradation of concrete. Finally, the project team visited 10 other utilities to conduct on-site investigations and develop photo documentation of impacts.

This final project report summarizes the most important research results from the work. The results of the 10 photograph intensive case studies are presented in a companion CD-ROM with associated descriptive text. However, the key results from the case studies have been synthesized below for sake of completeness. In total, ten key lessons were learned in the course of this work.

Lesson 1: The overall indication is that adverse impacts of enhanced coagulation on infrastructure do not always occur and are not catastrophic

In a survey of U.S. utilities, twenty-four percent reported noteworthy degradation of their treatment plant infrastructure due to coagulation, and 22% have had to resurface or repair concrete in their plant. Ten percent have observed signs of rebar corrosion in concrete, and this phenomenon might be more widespread because an additional 20% of utilities said they would not be able to recognize rebar corrosion problems. Other observed detrimental effects include: lower pH or higher dissolved CO_2 causing increased concrete degradation (6%), more failure of metal pipes, pumps, or structures in the plant (4%), more consumer complaints of red water

(2%), and increased copper pitting corrosion in consumer's homes (2%). Of course, given that design lifetimes of treatment plants can approach 100 years, and accelerated decay of that valuable asset may take decades to detect, the issue bears continued monitoring.

Lesson 2: Even if significant degradation of infrastructure occurs from enhanced coagulation, the costs at the treatment plant are not unbearable

Sixty-eight percent of utilities reported spending no money on infrastructure repairs due to enhanced coagulation, while 7% did not know how much was spent. Of the 25% of utilities who did spend money on repairs, the cost ranged from less than $1000 up to $100,000 per year, with an average of $13,000 per year. Nineteen plants reported annual costs exceeding $5000 each year, and if normalized on the basis of flow in MGD, the range of costs was $70-12,600 per MGD per year, with an average of $1850 per MGD per year. However, it is noted that utilities depreciate their assets differently, and there does not appear to be a generally accepted means of explicitly accounting for degradation.

Lesson 3: Utilities are anticipating problems and taking preventative measures to protect infrastructure

Twenty-five percent of utilities have applied epoxy or other protective coatings to protect concrete or metals, 13% are using cathodic protection systems for protection of metals, and 15% of utilities are considering using corrosion inhibitors to protect treatment plant materials from aggressive water (which may be complicated because some inhibitors are actually removed during coagulation). Case studies at Raleigh, Tampa and San Diego documented the benefits of epoxy and other concrete coatings in allowing effective coagulation even at pH less than 5 without losing infrastructure. Certainly, the rapid detection of a problem at San Diego, and then responding quickly with a protective coating system, probably averted major economic losses. However, as always, the quality of the job is important, as noted at Gothenburg where a poorly installed epoxy created major problems just a few years after installation.

Lesson 4: Concrete standards, materials and installation practice are important

Among the more startling findings, in some cases it appears that concrete structures installed after about 1960 have been much more susceptible to degradation from coagulation than older concrete. In many instances such as Gothenburg these comparisons have been made while treating virtually the same water source, leading to dramatic failures in just a few years. Australia provides a good example to follow in that they have specified strict standards that the water industry can employ to ensure an adequate product. Among the more important specifications, they require a water to cement ratio of 0.45, concrete coverage over rebar of > 6.5 cm and a standard for shrinkage of < 420 microstrains after 28 days. This is in addition to limits on chloride content of cement and development of blended cements more resistant to attack. In tours of plants across the U.S. and Canada, there is substantial shrinkage cracking in newer concrete. It was also noted that limestone versus siliceous aggregate led to marked differences in degradation mechanisms at Ottawa and elsewhere, but other than noting they were different, we cannot say which aggregate is better at this time. One hypothesis is that limestone aggregate dissolution acted to protect the mortar from attack by a sacrificial mechanism, whereas the converse argument is that loss of the aggregate exposes the mortar to more attack.

Lesson 5: The role of water quality in concrete degradation is key but poorly understood

When should a utility install protective coatings or install a type of concrete more resistant to attack? Even at the beginning of the project, it was well-known that the aggressiveness of a water to concrete is a function of temperature, mixing rate, dissolved CO_2, chloride, sulfate, pH, hardness, alkalinity, NOM, Zn^{+2} and other factors. However, when we looked at standard equations used to predict aggressiveness, they were not consistent with observations in the field, perhaps because the index ignores many key factors. Moreover, the case study at Ottawa demonstrated that some unusual films can form on concrete which might provide protection even in aggressive waters. At some plants, it is believed that ferric is more aggressive to concrete than alum, whereas in a direct test of coagulant aggressiveness in this work, alum turned out to cause much worse problems than ferric. These results indicate that we currently have no basis for informed decision making, and that fundamental research is necessary to sort out key factors. Until that work is completed, the main recommendation is that utilities carefully monitor impacts of coagulants, perhaps by installing concrete coupons and carefully inspecting basins. Learning to recognize signs of rebar degradation is important. However, if coagulation pHs at or below 6.0 are under consideration, it might be prudent to employ epoxy and or other protective coatings as a preventative measure.

Lesson 6: Free chlorine is highly corrosive

The notion that free chlorine is an oxidant and can corrode metals is well understood. However, few utilities anticipate increased problems from free chlorine when coagulation pH is lowered, even though such changes are known to enhance release of Cl_2 gas from water. Dramatic degradation was occurring in steel within treatment plants from this mechanism over a period of years, and even stainless steel was impacted to a degree. Painting of structures provides a simple means of slowing the rate of attack.

Lesson 7: Metallic infrastructure such as pumps and pipes bear close monitoring

In a case study at Billings, a very significant increase in frequency of copper pinhole failures occurred at the plant after switching from alum to ferric. While no one understands causes of pinhole leak failure, it appears that the change in coagulant made the water more susceptible to erosion corrosion, since the flow rate through the pipes stayed the same. Australian experience has demonstrated that epoxy coatings on pump bodies and impellers have reduced maintenance and pumping costs. Once again, U.S. utilities could probably benefit from this example.

Lesson 8: Coagulation at the treatment plant can cause discoloration and change corrosion of materials in the distribution system

Whenever a change in coagulant type or coagulation pH is implemented, the corrosivity of the water is fundamentally changed. If these changes are not countered at the end of the plant, they can cause problems in the distribution system. The experience of Lockwood is a good example of this, since a change from alum to ferric coagulant led to slight increases in consumer complaints of discolored water. The cause was traced to two sources including some ferric carryover through the plant and increased leaching of copper from home plumbing to water. Even small changes in the pH of distributed water can have noticeable impacts on corrosion of distribution system materials.

Lesson 9: If inhibitors are used to protect components in the treatment plant, they obviously need to be compatible with coagulation goals

Since phosphate or polyphosphate inhibitors are removed by, and likely interfere with coagulation, they do not appear to be a good option for protecting infrastructure. However, silicates are not removed by coagulants, and in some circumstances including low coagulation pH it is believed they occasionally improve aspects of coagulation. Likewise, Zn^{+2} dosing also appears to be compatible with goals of coagulation. We note at present that we do not have clear evidence that silicates or Zn^{+2} actually protect infrastructure, but merely that if they did do so they could be attractive options at utilities.

Lesson 10: Expect the unexpected

As we gain more experience with coagulant impacts over the next few decades, we can expect impacts on infrastructure that we cannot yet imagine. At Napa, for instance, improving the clarity of settled water by switching from alum to ferric led to increased algae growth in outdoor sedimentation basins, which in turn caused bubble formation and eventual destruction of a tube settler. Keep a close eye on your infrastructure, open your mind, and share your experiences with others.

CHAPTER 1
REVIEW OF POTENTIAL PROBLEMS ASSOCIATED WITH ENHANCED COAGULATION

There are many significant water quality changes that can occur during enhanced coagulation. In order to accomplish increased NOM removal, utilities are expected to either increase coagulant dose or switch from alum to an iron based coagulant, and many utilities may also increase acid addition (HCl or H_2SO_4) to lower the coagulation pH. These changes may lead to decreased NOM, lower pH, increased chloride, sulfate, calcium, and/or sodium, more iron or aluminum solids, more soluble aluminum, higher CO_2 levels, and decreased alkalinity. The effects of these water quality changes on various parts of water treatment plant infrastructure, including metal pipes, pumps, and tanks, as well as concrete basins are not well studied. In fact, the only reference found in the drinking water treatment literature was a case study in France, in which concrete losses of 0.12 inch (3 mm) were reported after ten years of alum coagulation (Giacasso 1988).

However, there are studies on metal corrosion in water distribution systems that are expected to be relevant to the behavior of metals in the treatment plant. In addition, there is ample information on degradation of concrete roads and bridges exposed to deicing salts (high Cl levels) or structures in marine environments (high Cl and SO_4), although the conditions are much more severe than those typically encountered in a water treatment plant. Nevertheless, this information has been reviewed to compile a list of potential problems that water utilities may encounter when enhanced coagulation is implemented (Table 1.1). Where appropriate, effects are listed for concrete, rebar, and metal components. Effects on lead and copper are also included to alert utilities of potential problems with distribution system piping.

DECREASE IN NATURAL ORGANIC MATTER (NOM)

Because the primary goal of enhanced coagulation is to remove more natural organic matter (NOM), utilities can expect lower NOM concentrations. NOM has been shown to be a component of protective films formed on concrete (AwwaRF and DVGW-TZW 1996) and iron (Larson 1966; Sontheimer, Kolle, and Snoeyink 1981; Campbell and Turner 1983; Edwards et al. 1994). Corrosion rates for copper are markedly decreased by the presence of NOM (Edwards et al. 1994), but corrosion by-product release is adversely impacted by NOM for copper (Dodrill and Edwards 1995; AwwaRF and DVGW-TZW 1996; Billings 1996; Korshin, Perry, and Ferguson 1996; Elfstrom Broo, Berghult, and Hedberg 1998; Korshin, Ferguson, and Lancaster 2000) and lead (AwwaRF and DVGW-TZW 1996; Kirmeyer et al. 2000; Korshin et al. 2000).

LOWER pH

pH levels will typically be depressed during enhanced coagulation due to increased coagulant dose or direct acid addition. Decreasing pH generally increases corrosion of lead, copper, and iron pipes (AwwaRF and DVGW-TZW 1996). Pore water in concrete is very alkaline, with pH as high as 13 to 13.5. If sufficient carbonate is present, a layer of calcium carbonate ($CaCO_3$) can form that seals the micropores on the concrete surface, which may protect the concrete from further degradation. If the pH is lowered, this calcite layer may dissolve (calcite solubility increases at lower pH) and the cement can disintegrate into the water (Legrand and Leroy 1990; AwwaRF and DVGW-TZW 1996). German standards recommend that the total inorganic carbon exceed 0.25 mM for satisfactory performance of concrete pipes; similarly, it has been proposed that a useful criterion for sufficient calcium carbonate formation potential in a water is to maintain a supersaturation ratio (Q/K) of 40 after corrections for ionic strength and temperature (AwwaRF and DVGW-TZW 1996). Langelier indices, calcium carbonate precipitation potential and other indices describe differing aspects of this same general hypothesis. However, it is commonly acknowledged that no single satisfactory index has been developed, or is likely to be developed, to describe concrete degradation in drinking waters (AwwaRF and DVGW-TZW 1996).

INCREASED CHLORIDE (Cl^-)

Utilities will see higher chloride levels if the dosage of a chloride-containing coagulant (e.g. PACl or $FeCl_3$) is increased or if HCl is used for pH adjustment. Chloride actually has little direct effect on concrete. However, the chloride can diffuse into reinforced concrete and contribute to corrosion of the iron rebar. The resulting scale (rust) has a much larger volume, which leads to cracking and spalling of the concrete (Figg 1983; Hope and Ip 1987; Hime 1994; AwwaRF and DVGW-TZW 1996).

There is some good information in the corrosion literature regarding recommended chloride levels for specific metal performance in drinking water applications. Galvanized materials can be expected to require maintenance every 10 years if chloride exceeds 200 mg/L. For type 304 stainless steel, it is recommended that maximum chloride not exceed 200 mg/L in Australia (Nicholas and Moore 1999), whereas a Swedish source cites less than 100 mg/L chloride as desirable (Alfonsson, Iverson, and Olsson 1999). Type 303 stainless is not recommended for water treatment plants by Australian experts (Nicholas and Moore 1999) and 316 stainless steels are likely to fail if chloride exceeds 400 mg/L or 200 mg/L based on Swedish and Dutch guidelines, respectively (Alfonsson et al. 1999). Mild steel has been found to require cathodic protection or epoxy for satisfactory service in critical applications. Higher chloride concentration has varying effects on distribution system pipes: it is beneficial for copper corrosion, detrimental to lead corrosion, and has mixed effects on iron corrosion (AwwaRF and DVGW-TZW 1996).

INCREASED SULFATE (SO_4^{-2})

Similar to chloride, utilities will see higher sulfate levels if the dosage of a sulfate-containing coagulant (e.g. alum or $Fe_2(SO_4)_3$) is increased or if H_2SO_4 is used for pH adjustment. Higher sulfate concentration has varying effects on metal pipes: it is detrimental to copper corrosion, beneficial for lead corrosion, and has mixed effects on iron corrosion (AwwaRF and DVGW-TZW 1996). Numerous studies have also been conducted on the adverse effects of sulfate on concrete and cement (Anonymous 1983; Figg 1983; Mehta 1986; Cohen and Bentur 1988; von Fay and Pierce 1989; Rasheeduzzafar et al. 1990; Anonymous 1991; Cohen and Mather 1991; Somuah et al. 1991; Grabowski et al. 1992; Tikalsky and Carrasquillo 1992; Hooton 1993; Tikalsky and Carrasquillo 1993; Dillon 1994; Al-Amoudi, Maslehuddin, and Saadi 1995; Day 1995; Freeman and Carrasquillo 1995; Mangat and Khatib 1995; Ghafoori and Zhang 1998; Janotka and Stevula 1998; Cannon 1999; Rollings, Burkes, and Rollings 1999;

Kurtis, Monteiro, and Madanat 2000; Mehta 2000). Most of the studies were for highways, bridges, and other structures exposed to seawater or marine atmospheres, so the sulfate concentrations were much higher (2–10%) than typically experienced in water treatment plants, although the mechanisms may be similar at lower sulfate levels.

There are two ways in which sulfate can lead to concrete degradation. First, sulfate in solution can react with the $Ca(OH)_2$ in concrete to form gypsum ($CaSO_4$), which is much softer and less durable. This form of attack generally occurs at higher (> 1000 mg/L) sulfate concentrations (Cohen and Mather 1991). Second, sulfate can react with the tricalcium aluminate (C_3A) present in cement to form the solid ettrengite ($3CaO{\bullet}Al_2O_3{\bullet}3CaSO_4{\bullet}32H_2O$) within the concrete pores:

$$3CaO{\bullet}Al_2O_3{\bullet}12H_2O + 3CaSO_4{\bullet}2H_2O + 13H_2O \rightarrow 3CaO{\bullet}Al_2O_3{\bullet}3CaSO_4{\bullet}32H_2O$$

Ettrengite has approximately double the volume of the original calcium aluminate, so its formation and crystallization results in internal pressure leading to cracks in the concrete (Figg 1983; Grabowski et al. 1992; AwwaRF and DVGW-TZW 1996). The amount of sulfate necessary to form ettrengite is not clear; one source reports that sulfate levels must exceed 500 mg/L before ettrengite will precipitate (Legrand and Leroy 1990) while another found that sulfate levels as low as 150 mg/L have been shown to damage concrete (Cohen and Mather 1991). Interestingly, if chloride in the water is high enough, as in the case of seawater, this type of degradation is not observed (Legrand and Leroy 1990; AwwaRF and DVGW-TZW 1996). Concrete exposed to a high sulfate solution at 50°C had twenty times more degradation than concrete at 22°C (Somuah et al. 1991).

MORE SOLIDS

The general idea that any deposit on concrete could be protective has been in the literature for more than 70 years (AwwaRF and DVGW-TZW 1996). Whether the surface is coated with iron, aluminum, calcium carbonate, or magnesium deposits, or organic matter, significant benefits have been reported. It is expected that enhanced coagulation will produce more solids, including $Al(OH)_3$ or $Fe(OH)_3$, due to higher coagulant doses. Several studies of cement-lined iron pipes in distribution systems described a brown layer containing high amounts of iron that acted as a protective barrier and prevented $Ca(OH)_2$ dissolution (Baylis 1927; Carson 1927; Chappell 1930; Flente and Sweitzer 1955; AwwaRF and DVGW-TZW 1996), so coagulation solids formed in the treatment plant may also be protective.

MORE SOLUBLE ALUMINUM (Al)

Although different solids may control aluminum solubility during coagulation, $Al(OH)_3$ is typically assumed to be the key solid formed. The pH of minimum solubility for $Al(OH)_3$ is between 5.7–6.0, so decreases in pH during enhanced coagulation may increase the amount of soluble aluminum in the water. This increased soluble aluminum may form other precipitates in the treatment plant or may later precipitate out in the distribution system. A few studies have indicated that aluminum solids may protect pipes from corrosion (Lauer, Lohman, and Rose 1999; Kirmeyer et al. 2000). However, a recent study found no evidence of this for lead and copper pipes (Kvech and Edwards 2001). The role of soluble aluminum in concrete and rebar corrosion has not been examined.

HIGHER CO_2/LOWER ALKALINITY

As utilities add higher doses of acidic ferric or alum coagulants, or add supplemental acids to enhance coagulation, the pH is depressed and alkalinity is consumed through well-established reactions:

$$HCO_3^- + H^+ \rightarrow H_2CO_3$$
$$H_2CO_3 \rightarrow CO_2\,(gas) + H_2O.$$

Thus, CO_2 concentrations increase and alkalinity decreases during enhanced coagulation.

Carbon dioxide is established as one of the most aggressive constituents to concrete, and it works synergistically with chloride to corrode rebar (Somuah et al. 1991; AwwaRF and DVGW-TZW 1996; Broomfield 2000). Concrete is subject to the carbonation process, whereby CO_2 diffuses into concrete and reacts with alkaline components to cause a gradual reduction in pH (Somuah et al. 1991; Mailvaganam 1992; Broomfield 2000; Glass, Reddy, and Buenfeld 2000; Krajci and Janotka 2000). The reaction with calcium hydroxide is most commonly considered as:

$$CO_{2(g)} + H_2O \rightarrow H_2CO_3$$
$$H_2CO_3 + Ca(OH)_2 \rightarrow CaCO_3 + 2H_2O.$$

However, one study found that $Ca(OH)_2$ was responsible for less than 25% of the acid neutralizing capacity of concrete, indicating that many other alkaline solid phases may also be important (Glass et al. 2000). Once the pH has dropped, the protective film on the iron rebar is destroyed and corrosion proceeds. Moreover, the decrease in pH can release chloride bound in Friedel's Salt ($C_3A{\cdot}CaCl_2{\cdot}10H_2O$), which further promotes the breakdown of the passive film (Rasheeduzzafar, Hussain, and Al-Saadoun 1992; Nagataki et al. 1993; Sanjuan 1997). Wet/dry cycling exacerbates carbonation (Anonymous 1995; AwwaRF and DVGW-TZW 1996; Broomfield 2000), so concrete basins may experience increased corrosion as water level fluctuates. For these reasons, German authorities recommend that CO_2 always be maintained at < 0.7 mM CO_2 in concrete conduits (Table 1.1) (AwwaRF and DVGW-TZW 1996), although no such guidelines have been established for concrete plant infrastructure.

CO_2 is also aggressive to most metals including iron and copper (AwwaRF and DVGW-TZW 1996; Edwards et al. 1996). Conversely, bicarbonate is among the most beneficial constituents to all materials (AwwaRF and DVGW-TZW 1996), with the exception of copper for limited circumstances (i.e., pitting, by-product release in low pH groundwater) (Edwards et al. 1996). Thus, the consumption of bicarbonate and production of CO_2 are doubly detrimental results of enhanced coagulation.

HIGHER CALCIUM/SODIUM

At many treatment plants, lime ($Ca(OH)_2$) or caustic (NaOH) is added to water in conjunction with coagulation to maintain the pH above certain target values. While this may seem to be in direct conflict with goals of enhanced coagulation, it is sometimes necessary to maintain effective particle removal (Tseng, Segal, and Edwards 2000). Moreover, if the caustic or lime is properly dosed, the improvements in TOC removal from the extra $Fe(OH)_3$ and $Al(OH)_3$ can overcome reduced TOC removal from slightly higher coagulation pHs (Edwards 1997; Tseng and Edwards 1999). Thus, calcium and sodium levels might increase in the water of some utilities practicing enhanced coagulation. Sodium seems to be relatively inert to materials, whereas calcium has well established benefits, particularly for concrete (AwwaRF and DVGW-TZW 1996). German experts recommend that calcium levels be > 0.02 mM in waters with concrete conduits (Table 1.1) (AwwaRF and DVGW-TZW 1996).

Table 1.1
Water Quality Changes from Coagulation and Possible Impacts on Materials Degradation

Water Quality Change	Reason for Change	Potential Consequence For...		
		Concrete	Rebar	Metals
Lower NOM	Removed by Coagulation	Adverse	Unknown	Adverse
Lower pH	Higher coagulant doses and direct acid addition	Severe adverse	Adverse	Severe adverse
More Chloride	Higher doses of ferric chloride, polyaluminum chloride or HCl	Neutral, very high levels may	Severe	Variable
More Sulfate	Higher doses of alum, ferric sulfate, or sulfuric acid	Adverse for some types of	None	No Effect or Detrimental
More $Al(OH)_3$ or $Fe(OH)_3$ solids	Typical consequence of higher coagulant doses	Benefit	Unknown	Unknown
More soluble Al^{+3}	Higher alum doses and lower pH may increase solubility of solids formed	Possible benefit	Unknown	Unknown
Higher CO_2	Acidification of waters containing bicarbonate	Severe adverse; DVGW	Adverse	Adverse
More Calcium	Addition of supplemental lime to maintain target coagulation pH and	Benefit; DVWG recommen	Unknown	Some benefits

A FOCUS ON MATERIALS IN RELATION TO DEGRADATION

There is no single material that possesses the ideal combination of corrosion resistance, price and durability for all applications in water treatment plant infrastructure. However, there are factors to be considered in selecting appropriate materials for the anticipated water quality when doing repairs or building new treatment plant infrastructure. The following sections briefly describe some of these issues for key infrastructure materials (concrete and rebar) and two rehabilitation techniques (cathodic protection and protective coatings).

Concrete

In addition to the water quality criteria cited earlier (e.g., recommended total inorganic carbon, calcium, and dissolved CO_2 levels), the type of cement, curing conditions, and other factors related to exposure are also very important in determining the durability of concrete. There are many types of Portland cement, with variable strengths and chemical resistance that can be included in concrete (Kosmatka and Panarese 1992). As a general rule, it is recommended that water treatment plants use a high durability aggregate, a water to cement ratio < 0.45, and a water cure (Ellis and Moore 1993; Nicholas and Moore 1999). Outside of these references, there is a general lack of information in the literature for making concrete used in water treatment plants resistant to degradation. Although much of the available information about concrete has been developed either for concrete in distribution systems (which may not reflect the mechanisms of corrosion as they occur for concrete in treatment plants), or for transportation infrastructure (where concrete is exposed to more severe conditions than the typical water treatment plant), this information may be useful for drinking water applications.

Chloride Resistance

Nearly 80 publications were found that describe the relationship between concrete composition and environmental factors on chloride penetration and resulting rebar corrosion. However, nearly all of these studies were on highways, bridges, and other structures exposed to deicing salts or marine environments, so the chloride concentrations were much higher (3 – 4% NaCl) than levels typically experienced in water treatment plants. Moreover, there are many types of cement composition and exposure conditions used in the various studies. However, several trends are clear. In general, concrete with lower permeability will be more resistant to chloride attack because chloride diffusion into the concrete is limited. The proper water to cement ratio is critical for making a low permeability concrete (Sivasundaram, Carette, and Malhotra 1991; Johnston 1994; Baweja, Roper, and Sirivivatnanon 1998; Baweja, Roper, and Sirivivatnanon 1999; Detwiler, Whiting, and Lagergren 1999; Gu et al. 1999; Gu et al. 2000). The temperature at which concrete is cured may also play a role in its chloride resistance. Concrete cured at higher temperature is more permeable to chloride (Detwiler, Kjellsen, and Gjorv 1991; Sanjuan 1999), although the use of silica fume or blast furnace slag decreased chloride permeability at these higher temperatures (Detwiler, Fapohunda, and Natale 1994).

Chloride diffusion or permeability can also be reduced by including a pozzolanic material in the concrete such as silica fume (Figg 1983; Berke and Hicks 1994; Johnston 1994; Detwiler et al. 1999; Gu et al. 2000; Hou and Chung 2000; Wee, Suryavanshi, and Tin 2000), blast furnace slag (Figg 1983; Baweja et al. 1998; Gu et al. 2000; Wee et al. 2000), latex and carbon fibers (Hou and Chung 2000) and microsilica (Li, Peng, and Ma 1999). Fly ash, a byproduct of coal combustion, can also be used to replace some of the Portland cement, which will form a calcium silicate hydrate that fills some of the voids in the concrete, which should reduce permeability. Most studies report a benefit from adding fly ash, with a decrease in chloride permeability (Sivasundaram et al. 1991; Johnston 1994; Alhozaimy, Soroushian, and Mirza 1996; Scanlon and Sherman 1996; Baweja et al. 1998; Gu et al. 1999; Li et al. 1999; Malhotra,

Zhang, and Leaman 2000) or free chloride ions (Haque, Kayyali, and Gopalan 1992; Haque and Kayyali 1995). However, one study reported more carbonation and increased cracking from chloride induced corrosion when fly ash was added (Haque and Kawamura 1992) and another study found more hydrated, less protective iron oxide scales on rebar when fly ash was present (Montemor, Simoes, and Ferreira 1998).

Corrosion inhibiting admixtures or curing compounds including calcium nitrite and organic–, boron–, or phosphorus–based compounds have been successfully used to decrease chloride permeability of concrete (Loto 1992; Nmai, Farrington, and Bobrowski 1992; Collins, Weyers, and Al-Qadi 1993; Gu et al. 1997; Saricimen et al. 1998; Zhang et al. 1999), although several studies reported the opposite result, with increased chloride permeability when calcium nitrite inhibitor was used (Li et al. 1999; Loulizi Al-Qadi and Diefenderfer 2000). Moreover, there is also some controversy regarding the validity of the Rapid Chloride Permeability Test (the standard test used to evaluate chloride penetration into concrete) (Schupack 1989; Bishara 1991; Luping and Nilsson 1992; Haque and Kayyali 1995; Khan 1997; Shi, Stegemann, and Caldwell 1998) especially when these admixtures are used (Shi et al. 1998; Wee et al. 2000a), so results based on this assessment test should be interpreted with caution.

Finally, a rehabilitation method for bridges and road decks suffering from chloride contamination has been developed wherein the chloride is extracted from contaminated concrete by applying an electric field (Glass, Zhang, and Buenfeld 1995; Velivasakis, Henriksen, and Whitmore 1997; Velivasakis, Henriksen, and Whitmore 1998; Castellote, Andrade, and Alonso 1999). It is postulated that this electrochemical treatment will also cause a layer of $Ca(OH)_2$ to precipitate at the rebar surface, which will resist future decreases in pH and increase the corrosion resistance of the rebar (Glass and Buenfeld 2000). However, this process takes 4 – 8 weeks and may not be practical for water utilities.

Sulfate Resistance

In general, concrete with lower permeability will also be more resistant to sulfate attack because sulfate diffusion into the concrete is limited. This can be accomplished by using a lower water to cement ratio, curing the concrete completely, or using air-entraining admixtures (Day 1995; Ghafoori and Zhang 1998). Similar to the case of chloride resistance, many pozzolanic materials such as fly ash (Figg 1983; Mehta 1986; von Fay and Pierce 1989; Tikalsky and Carrasquillo 1992; Tikalsky and Carrasquillo 1993; Al-Amoudi et al. 1995; Freeman and Carrasquillo 1995; Mangat and Khatib 1995; Ghafoori and Zhang 1998), silica fume (Figg 1983; Cohen and Bentur 1988; Rasheeduzzafar et al. 1990; Hooton 1993; Al-Amoudi et al. 1995; Mangat and Khatib 1995; Wee et al. 2000a), blast furnace slag (Mangat and Khatib 1995; Wee et al. 2000b), or zeolite (Janotka and Stevula 1998) also decrease permeability and increase sulfate resistance. Finally, special types of cement including high alumina cement or low C_3A cement (sulfate resisting Portland cement, SRPC) (Figg 1983) are available. For example, Type II and Type 5 Portland cements are recommended for higher sulfate concentrations from 150-1500 mg/L and above, respectively. There are also waterproof cements and plastic-type cements that alter sulfate resistance (Kosmatka and Panarese 1992).

Rebar

In general, the rate of rebar corrosion is proportional to the chloride level. Given that it is encased in concrete, rebar is often best protected by the same things that protect concrete including sealers, coatings, membranes, or polymers (Ellis and Moore 1993; Weyers 1993; Nicholas and Moore 1999) or by using the special blends discussed above to decrease the chloride permeability or increase sulfate resistance. In structural or highway applications, rebar that is galvanized, stainless steel clad, epoxy-coated or made from different alloys has been successfully used (Rasheeduzzafar et al. 1992; Yeomans 1994; Alonso et al. 1998; Montemor et al. 1998; Saricimen et al. 1998; Pyc et al. 2000; Trejo et al. 2000). Zinc-rich coatings were also found to provide some cathodic protection to rebar in saline solutions, although the coatings did not provide a barrier type protection (Morris, Vazquez, and De Sanchez 2000). It must be emphasized that any holidays in these coatings will be the site of accelerated corrosion that can proceed under the coating (Yeomans 1994). Moreover, in water treatment plants, the exposure environment becomes very important since concrete that is below, near, or well above the water line will have differences in rates of attack.

CATHODIC PROTECTION

If the water quality is expected to be corrosive to various existing metal elements in a water plant infrastructure, the metals may be protected by cathodic protection (CP) with either an impressed current or a sacrificial anode of graphite, zinc or magnesium. CP has been successfully used to protect metal water storage tanks, iron pipes, and clarifier rake arms (Sudrabin, Kaiser, and Liszczynsky 1983; Thompson, Lawson, and Beavers 1987; Benedict et al. 1997; Gummow 2000). CP has also been used for prestressed concrete cylinder pipe (Benedict 1990; Benedict et al. 1997) although another study reports that cathodic protection should not be used on prestressed concrete because of concern that hydrogen embrittlement of the high-strength steel may occur (Paul 1998). Moreover, CP may not be appropriate for concrete that contains epoxy-coated rebar due to concerns about electrical continuity. The ANSI/AWWA Standard D104-97 ("Automatically Controlled, Impressed-Current Cathodic Protection for the Interior of Steel Water Tanks") provides guidelines for cathodic protection of steel tanks.

COATINGS

Many types of coatings for both metal and concrete are available, including bituminous, rubber, acrylic, epoxy, polyurethanes, silicones, and silanes. In addition to selecting the appropriate coating for the expected water quality, coatings used in drinking water facilities must also be certified safe for contact with potable water according to NSF 61 standards (Table 1.2). Other environmental concerns include air pollution due to volatile organic chemicals (VOCs) from the solvents used to apply the coating (although solventless epoxies are becoming more popular), dust generation (from blast cleaning to prepare the surface prior to applying coating or to remove previous coatings) and possible leaching of organic chemicals from coating into water (Harper 1991; Anonymous 1997a).

Coatings for Concrete

It is the German belief that for pipes, epoxy is not cost effective relative to simple replacement of the concrete (AwwaRF and DVGW-TZW 1996). However, the cost–benefit analysis is likely to be dramatically different in treatment plants than in the distribution system. For instance, at treatment plants coatings are easier to apply, they can be readily inspected, and the surface area of material protected per unit volume of water present is lower. In addition, waters are much more aggressive in the treatment plant under conditions of enhanced coagulation, since pH is commonly raised before the distribution system for corrosion control. All of these differences increase the relative attractiveness of coatings.

A few case studies are presented in the literature on the use of coatings to protect concrete components in water treatment plants. Epoxy has been successfully used on various concrete basins in a water treatment plant including raw water receiving wells, flocculation basins, sedimentation basins, clearwells, and distribution reservoirs (Anonymous 1997b; Gotoh and Shirozu 1997) and a urethane lining with epoxy primer was used to protect a precipitator (Anonymous 1995). Surface coatings including epoxy and cement-based polymer modified repair mortar have also been successful for repairing various concrete structures in water treatment plants (Groves 1999).

There are also case studies of successful use of coatings to protect concrete in other industries besides water treatment (Munger 1984; Foscante and Kline 1988; Munger 1990; Mailvaganam 1992; Greenfield 1994; Ilaria 1995; Bierwagen 1998; Luma 1998; McGovern 2000). Various silanes, epoxies, polyisocyanate, silicon resins, and acrylic based surface coating have been used to prevent chloride diffusion or carbonation of concrete in the transportation infrastructure (Whiting and Nagi 1996; Johnson and Skerry 1998; Suryavanshi, Swamy, and McHugh 1998; Swamy, Suryavanshi, and Tanikawa 1998; Al-Gahtani et al. 1999), although sodium silicate sealers were not effective (Whiting and Nagi 1996; Al-Gahtani et al. 1999). Several studies have also examined the use of epoxy, cement, or other materials as protective coatings in sewer systems or wastewater treatment plants (Kline 1983; Szymanski and Scott 1983; Smith 1989; Reyna, Vanegas, and Khan 1994; Steele 1994; McGovern 1999), and another study found that addition of silica fume and organic corrosion inhibiting admixtures to the concrete used for sewer pipes made it less susceptible to corrosion (Daczko, Johnson, and Amey 1997). Coatings have also been used on salt water tanks at an aquarium (Carr 1991), the wastewater basins at a metals finishing plant (Anonymous 1992), and concrete secondary containment structures for nuclear and chemical facilities (Chmelar 1992; Finch 1995; Davis 2000).

Coatings for Metals

The Engineering and Water Supply Department of South Australia has conducted a thorough evaluation of different materials and protection options for steel components in water treatment plants (Ellis and Moore 1993). In those studies, flake reinforced polyester coatings performed best, but had some problems associated with lack of adhesion. Epoxies were found to be the next best option for coating steel. Coal tar epoxy and micaceous iron oxide epoxy were favored because they are economical and easy to apply, and high build epoxies performed better than solventless epoxies. Vinyl coatings also worked well, but required multiple coats so application was deemed uneconomical. Inorganic zinc silicates and rust treatments performed well in atmospheric environments but not underwater.

Epoxy linings have also been successfully used for metal potable water storage tanks and iron water mains (Anonymous 1994; Bromley 1995; Wernham 1995), and epoxy coatings were shown to protect cast iron, bronze, and stainless steel pump components (Ellis and Moore 1993). A thermoplastic polymer has also been used for water treatment plant pipes, fittings, and pumps (Pey 1997). Epoxy, vinyl, and chlorinated rubber linings for water storage tanks have also been used in conjunction with cathodic protection (Sudrabin et al. 1983).

Again it is emphasized that the coating must be applied properly to ensure good performance. Gaps in the coating ("holidays") can actually lead to increased corrosion at the holiday site. An additional problem that has been encountered is blistering of the epoxy lining applied to a potable water tank due to evaporation of the water soluble solvent from the epoxy primer (Weldon 2000). Recommendations to avoid this problem include applying the coating in warmer weather or using forced–air ventilation to allow the solvent to evaporate. Additional guidance for application of coatings can be found in the Standards listed in Table 1.2.

Table 1.2
Applicable US standards for coatings applied to water plant infrastructure

Standard*	Description
NSF 61	Drinking Water System Components in Contact with Drinking Water
ANSI/AWWA C116/A21.16-98	Protective Fusion-Bonded Epoxy Coatings for the Interior and Exterior Surfaces of Ductile-Iron and Gray-Iron Fittings for Water Supply Service
ANSI/AWWA C210-97	Liquid-Epoxy Coating Systems for the Interior and Exterior of Special Sections, Connections, and Fittings for Steel Water Pipelines
ANSI/AWWA C213-96	Fusion-Bonded Epoxy Coating for the Interior and Exterior of Steel Water Pipelines
ANSI/AWWA D102-97	Coating Steel Water-Storage Tanks
ANSI/AWWA D130-96	Flexible-Membrane-Lining and Floating-Cover Materials for Potable-Water Storage
ACI 515.1R-79	A Guide to the Use of Waterproofing, Dampproofing, Protective, and Decorative Barrier Systems for Concrete

* NSF = National Sanitation Foundation
ANSI = American National Standards Institute
AWWA = American Water Works Association
ACI = American Concrete Institute

CHAPTER 2
EXPERIENCE GAINED BY THE WATER INDUSTRY: METROPOLITAN WATER DISTRICT, EXPERTS FROM AROUND THE WORLD, AND A UTILITY SURVEY

This Chapter is organized into three sections reporting direct experiences of the water industry. The first section summarizes important experiences obtained at the Metropolitan Water District of Southern California. The second describes inputs and publications obtained from experts in Australia, England and France. The final section presents data obtained from a survey of coagulation utilities in the United States.

METROPOLITAN WATER DISTRICT OF SOUTHERN CALIFORNIA

The Metropolitan Water District (MWD) of Southern California has obtained considerable experience with planning for protection of infrastructure from enhanced coagulation, and laboratory study results regarding resistance of treatment plant materials to concentrated or dilute alum and ferric. At the time of the site visit, the plant was planning to construct two new pre-treatment facilities for ozonation after depressing pH to 6.0 using sulfuric acid. After this ozonation step, a combination of coagulant and NaOH will be dosed to the water to raise pH to 7.0 throughout existing conventional treatment facilities.

Paper Study to Design New Pretreatment Facilities and Protect Existing Infrastructure

Bill Sleeper and the infrastructure group at MWD attempted to anticipate detrimental effects of low pH of concrete degradation. In the course of their study, they found the following guidelines from the literature (Table 2.1) (Alexander et al. 1994).

Table 2.1.
Recommended limits for assessing aggressiveness of water to concrete (from Alexander et al. 1994).

Parameter	Degree of Aggressiveness			
	Moderate	High	Very High	Excessive
pH	6-8	5-6	4.5-5	< 4.5
ΔpH*	-0.2 to –0.3	-0.3 to –0.4	-0.4 to –0.5	< -0.5
Calcium (mg/L)	80-120	40-80	20-40	< 20
Sulfate (mg/L)	150-1000	1000-2000	2000-3000	> 3000
Cl- (mg/L)	500-1000	1000-2500	2500-5000	> 5000

* ΔpH = Actual water pH – pH of calcite saturation

The paper also puts forward the concept of Leaching Corrosion Index (LCI) to gauge the combined effect of various chemical factors on concrete:

$$LCI = 0.333[200(9.5\text{-pH}) - 2000(\text{pH}) + 5.5(200\text{-Ca}^{+2})]$$

The LCI is corrected for flow if the water is not exposed to laminar flow conditions. If flow is turbulent the LCI value is multiplied by 1.75, but if flow is stagnant LCI is multiplied by 0.5. Finally, the LCI is corrected for temperature by multiplying by the factor [1+(0.05(T-20))], which yields a higher LCI if temperature is above 20°C and a lower LCI if temperature is below 20°C.

As a general rule, if the LCI is below 350, a cement with water to cement ratio less than 0.45 should be long-lived without protective coatings. If LCI is 350–750, a higher grade cement might be recommended, including a 50:50 mixture of Portland Cement with blast furnace slag, a 70:30 mixture of Portland Cement with fly ash, or a 90:10 mixture of Portland Cement with condensed silica fume. If the LCI is above 750, a protective coating for the concrete is recommended.

Applying this index to MWD data for August 1994 at Silverwood Lake when calcite saturation pH was 8.11, influent pH = 8.12, and Ca^{+2} = 24 mg/L yielded an initial LCI of 408. After correcting for an actual temperature of 24°C and turbulent conditions in the rapid mix, the LCI increased to 857, which suggested a problem with the water to start with. Decreasing the pH to 6.0 in this same water using H_2SO_4 leads to an LCI of 4121. This was noted to exceed the databases used by Alexander et al. to develop the concrete corrosion indices, and overall it seems that this approach is not calibrated very well to treatment plants since it suggests most facilities require coatings. However, such an approach would be extremely valuable to protection of treatment plants if it were to be calibrated. Nonetheless, the approach did anticipate a significant problem with concrete at pH 6.0.

Since the MWD standard concrete mix exceeds the recommendations of Alexander et al. (1994), and it was believed that ozone would degrade any organic coatings that might be applied, a concrete coating approach was recommended for the basin exposed to pH 6.0 water during pre-ozonation (Sleeper, 1995). In practice, MWD eventually decided to construct the entire pre-treatment facility with silica fume concrete. This higher strength concrete allowed significant reductions in the mass of the concrete facilities poured, as well as providing a greater resistance to acid conditions.

The next question to be addressed was “What should be done about protecting the existing infrastructure where coagulation takes place?” MWD carefully considered either maintaining a minimum coagulation pH of 7.0 or coating the basins. The protective concrete coatings that were considered included 1) epoxies such as coal tar, Novalac and amine cured, 2) a mixture of sand, Portland cement and 35% silica fume known to have extremely low permeability that can be spray–applied to 1” thickness, and 3) a potassium silicate coating applied in a multi-layer system including a base layer, impermeable matting, and a final layer of potassium silicate material. The epoxy requires at least 3 days cure time whereas the cements require about 1 week to cure. Catalyst leaching to water from the epoxy is a major concern, since significant leaching is known to occur for up to one month and initial water contacting the material might need to be discarded, but leaching is not known to be a problem for the cementatious type coatings. For the 493,023 square feet area to be protected, total estimated costs were $2-2.7 million for the epoxy approaches, versus $9.4 and $13.3 million for the 35% silica fume and potassium silicate approaches, respectively.

In the end, since they believed that any concrete coating technique would require maintenance on 3-5 year intervals with an overall expected lifetime of 10-15 years, the coating approach seemed less desirable than maintaining a coagulation pH of 7 given the inconvenience of shutting the plant down. It is possible that as the water quality group considers whether lower pH coagulation will be required in the existing conventional treatment train, the materials group would reconsider whether coatings could be used to protect existing infrastructure. At present maintaining coagulation pH 7 seems to be a reasonable compromise.

All metallic components at the plant were already coated or protected cathodically and were therefore not a focus of thc study

Laboratory Testing of Ferric Degradation

At the time ferric chloride dosing was also being considered, so MWD conducted some long term materials testing using both concentrated (42%) and dilute (1.9 mg/L ferric chloride) dosing. Table 2.2 summarizes results of the testing. Tests with concentrated ferric chloride led to specifying Titanium alloy grade 7 or Hastelloy C-22 for metallic components contacting the ferric and for mechanical diffusers of the chemical.

Laboratory testing with dilute ferric solutions showed that most metals were adversely impacted by a low level of ferric when compared to a low level of alum. An additional concern was that ferric actually passivated zinc anodes, which could cause problems if zinc was used sacrificially to protect other metals.

Finally, some laboratory testing of Portland cement coupons was also compared at a pH of 6.3 versus normal pH of 8.0 for a six month exposure period. The low pH specimens did experience significant weight loss compared to the control, and were also subject to greater water uptake and more surface deposits that could be removed by light brushing. Total coupon weight loss was about 1% after exposure, drying and cleaning at the lower pH versus a weight gain of about 0.1% at the higher pH condition. Interestingly, one study exposing concrete to concentrated ferric chloride resulted in depth losses of only ¼" in a year. This seems to be quite low given that San Diego noted losses of about ¼" concrete in their coagulation water at pH 6.6, and points either to the importance of concrete in resisting attack or perhaps that a mechanism other than low pH is important in controlling attack rates.

Table 2.2.
Summary of MWD testing results on materials performance after 1 year exposure to concentrated ferric coagulant in tests at La Verne Materials lab

Not acceptable = corrosion rate above 50 mpy, poor = corrosion rate between 10 and 50 mpy, moderate = corrosion rate between 2-10 mpy, very good = corrosion rate less than 2 mpy. Elastomer performance based on various parameters. Results for $FeCl_3$ (average dose = 1.9 mg/L) versus alum (average dose = 5.3 mg/L) based on 1 year head to head tests conducted in the flocculator at Mills filtration plant, with median pH of 8.0.

Material	1 Year results with concentrated $FeCl_3$	$FeCl_3$ versus alum
Copper/ Copper Alloys	Not acceptable	All corroded at less than 1 mpy, but ferric caused 4-8X the corrosion of alum
Carbon and Steel	Not acceptable	Ferric more aggressive than alum, and corrosion rates > 2 mpy
Aluminum alloys		Alum more aggressive than ferric, and corrosion rates < 2 mpy
Stainless: 304, 316, 317	Not acceptable	Very low corrosion rates, welded areas did worse for both coagulants
Cobalt Alloys	Very good: Elgiloy and MP-35-N Not acceptable: Stellite 6B	
Zinc		Ferric twice as aggressive as alum, suggesting problems for galvanized materials
Magnesium		Alum more aggressive than ferric, suggesting ferric might cause problems with sacrificial effect.
Nickel Alloys	Very Good: Hastelloy C-22 and C-276 Poor: Hastelloy X Not Acceptable: Monel 400 and Hastelloy G	
Zirconium Alloys	Poor or not acceptable: Zircadyne 702, 704 and 705	
Titanium Alloys	Very Good: TiGr 2, 4, 5, 6, 7, 11, 12 and Ti Beta C	
Duriron	Moderate-Very Good: Durichlor 51	
Elastomer	Very good: Neoprene, Natural Gum, Viton Moderate: Hypalon, silicone sheet stock, Buna-N	
Plastics	Very good: Polypropylene, LDPE, UHMW Discrepancy: PVC and Hi-Polymer Vinyl Chloride cited as very good in Test 1, but PVC and vinyl listed as poor in more thorough Test 2.	
Sealants	Not acceptable or Moderate: Polysulfide base T227, polyurethane base elastothane 227 Very good: Dow Silicone Base Caulking	No significant difference between alum and ferric. Polysulfide sealants generally unacceptable.
Concrete Coatings	All maintained coating integrity including modified polyester, modified vinyl ester with sand, amine and high solids and polyamide epoxys, modified vinyl ester, neogard poolguard II, fluid polymers HMP 45.5	No significant differences between alum and ferric.
Fiberglass Pipe or Coatings	No significant degradation.	

INTERNATIONAL PERSPECTIVES ON INFRASTRUCTURE DEGRADATION

Opinion was sought from experts in the United Kingdom (UK), Australia and France to define experiences in infrastructure degradation from coagulation in each country.

Infrastructure Degradation from Coagulation in the UK

Alum is the traditional coagulant of choice at most UK water treatment plants, although ferrous sulfate, ferric sulfate and ferric chloride receive occasional use and have become increasingly popular since the 1970's as manganese-free ferric became available and other quality improvements were made. Polyaluminium chloride (PACl) became available at an acceptable price in the early 1980's concurrent with the introduction of liquid alum. Cationic organic polymers receive some limited use.

Color in water supplies varied markedly as did treatment effectiveness in the aftermath of World War II. Treatment has improved continuously ever since, most notably with adoption of EC water quality standards demonstrating a link between consumer complaints and influent water quality to the distribution system. In particular, maintenance of iron and aluminum residuals of less than 100 µg/l was recommended, which was more formally accepted in the Water Quality Regulations of 1989 which mandated less than 200 µg/l Fe or Al at customers' taps. With such stringent standards on metal residuals and the desire to improve removal of organics, lower pH coagulation has become increasingly common and ferric is favored in such circumstances. In higher alkalinity waters, acid is also occasionally used, although plant operators are still biased towards minimizing chemical doses.

Problems with corrosion are perceived to be important only in low alkalinity waters when treatment pH drops below 6.0, and the pH can drop as low as 4.8 during coagulation of highly colored surface water. The concrete at many treatment plants is showing eroded cement and exposed aggregate. Mild steel components were originally heavily galvanized, but stainless steel has become increasingly popular for flocculator impellers, recycle pipework and sludge scraping mechanisms in low pH waters. For the smaller plants, mild steel tanks with robust linings can be economically justified. Corrosion–resistant concrete is used, but there are reports of resistance not being what it should or pH that is too low. In some cases, protective coatings have been applied.

Corrosion that distinctly affects asset life is most likely to be localized, such as where acidic chemicals briefly result in low pH in the vicinity of dosing before they become dispersed and diluted. Poorly engineered or maintained chemical handling and dosing facilities are more likely to result in corrosion of structures and other equipment and can lead to expensive maintenance or refurbishment. Moreover, substantial costs may be incurred if low pH water treatment is pursued in plants with such inappropriate structural materials.

Infrastructure Degradation from Coagulation in Australia

David Nicholas, Greg Moore and Brian Ellis have documented Australian experiences with infrastructure degradation from coagulation at treatment plants. These investigators were interviewed during a trip to Australia, and they also filled out the same survey mailed to U.S.

utilities. In combination these investigators were able to give highlights of experiences at 12 drinking water and 20 wastewater treatment plants.

The Australian utilities surveyed all use alum coagulants and the issue of degradation is not a major consideration. They are currently moving towards higher coagulant doses and lower coagulation pHs, and while concrete degradation is a concern, they have yet to see significant signs of accelerated attack. Corrosion inhibitors such as silicates and phosphates are not used, nor is future use under consideration. Some trials are underway for use of protective coatings to concrete.

One possible reason that concrete degradation is not currently a more serious problem is that a Concrete Deterioration Committee was formed in September 1983 in response to problems in the late 1970s. The committee prepared a final report in 1984, outlining standards and practices for new concrete and effectiveness of surface repair products. These recommendations were proven valid in later field tests at a water treatment plant. Specifically, the key changes called for reducing the water to cement ratio from 0.55 to 0.45 and the coverage over reinforcement was increased from 4 cm to 6.5–7.5 cm on all waters–retaining structures. A new standard for shrinkage of < 420 microstrains after 28 days was also instituted, whereas shrinkage was previously unspecified. Likewise, new limits were set on the chloride content of each individual ingredient (water < 0.05%, admixture < 0.2%, cement < 0.01%, aggregate < 0.01% weight/volume), and blended cements were developed that are particularly resistant to chloride attack of rebar in water treatment plants. Institution of the new standards has led to an obvious and continual improvement in concrete performance with time.

Interestingly, according to Nicholas, erosion corrosion and stress corrosion cracking (SCC) are only rarely encountered at drinking water treatment plants. It is recommended, of course, that galvanic couples be avoided to the extent possible. With respect to specific metals, mild steel has been found to require cathodic protection or epoxy coating for satisfactory service in critical applications. The performance of products within each generic group of protective coatings can vary markedly, with factors of 2 difference in performance not uncommon. Recent testing of solvent free epoxy is showing encouraging results in potable water applications. Galvanized materials can be expected to require maintenance every 10 years if chloride exceeds 200 mg/L. For 304 stainless steel, it is recommended that chloride not exceed 200 mg/L. Type 303 stainless steel is not recommended for water treatment plants.

Substantial advances have also been made in terms of extending the performance, useful life and time between major overhauls in pumps (Ellis and Moore 1993). Specifically, use of high build epoxies in the body of pumps on the impellers have improved the pumping efficiency to equal or better than that of the original pump, also producing additional marked savings in electricity costs. Ceramic–filled epoxies have especially high abrasion and wear resistance, making them particularly well suited to this application.

Infrastructure Degradation from Coagulation in France

Several reports in relation to cement infrastructure degradation of potable water tanks or filters were forwarded to the author by Dr. Pierre Leroy, and these reports are also described elsewhere (AwwaRF and DVGW-TZW, 1996). A few points relevant to this project are summarized below. First, it is believed that natural organic matter protects concrete from degradation. Since a goal of enhanced coagulation is to remove NOM, increased aggressiveness of water after higher coagulation doses might be directly due to NOM removal, as opposed to higher pH, SO_4^{-2} and Cl^-.

Another interesting observation was made in a Paris reservoir related to the location of concrete degradation. In that study, it was observed that concrete in deeper water "carbonates" about 5 times more slowly than at the water surface. Not enough information was gathered to determine if the site of degradation at upper levels was below or above the water line, and due to varying water levels such differentiation might not be possible. Nonetheless, it points to the possibility that degradation can be depth dependent, and utility monitoring activities should take this into account.

A more recent publication by French researchers (Soukatchoff, Baron, and Lessirard 2001) presents data demonstrating that aluminum leaching to water from dissolved cement can be greatly reduced by the presence of silica in the water. This points to the possibility of adding silica as a concrete corrosion inhibitor in some waters when silica concentrations are naturally low. Experiences of these researchers with cement pipe linings exposed to soft low pH water has indicated that resistance to degradation decreases in the order: High Alumina Cement > Blast Furnace Slag Cement > Portland Cement. Lifetimes of greater than 100 years can be expected for high alumina cement liners and lifetimes of greater than 50 years can be anticipated for blast furnace slag liners. However, high alumina cement received much negative publicity in the 1970's when several buildings constructed with the product collapsed, and its use in many applications has been banned in some countries. It is uncertain what effect these documented deficiencies for high alumina cement will have on its suitability as a pipe lining material.

Acknowledgements

For the section on infrastructure degradation by corrosion in the UK (by Ross Gregory), there was a useful dialogue with representatives of North of Scotland Water, North West Water, Severn Trent Water, South West Water, Welsh Water and Yorkshire Water.

UTILITY SURVEY ON IMPACTS OF ENHANCED COAGULATION

Survey Implementation

In April 2001, a five-page survey (see Appendix A) was mailed to all surface water treatment utilities in the AWWA WaterStats database that reported practicing coagulation. A follow-up reminder postcard was mailed in May to utilities who had not yet sent their surveys back. A total of 381 surveys were mailed out, and 7 were returned with undeliverable addresses. The survey response rate was 60%, with 225 utilities responding. Twenty-nine of the participating utilities had multiple treatment plants and therefore returned more than one survey, so a total of 275 surveys were received. Forty-eight utilities (13%) also sent water quality reports. Utilities from 45 of the 50 US states are represented in this survey. The AWWA database does not include any utilities from Mississippi, New Mexico, or North Dakota, and contains only one each in Delaware and Vermont, but those utilities did not respond. Nearly 99% of the utilities responding to the survey were correctly identified as practicing coagulation. It should be noted that the survey methodology employed in this work does not allow calculation of statistical confidence levels for differences in answers. However, it is still useful to discuss general trends in the paragraphs that follow.

Utility Characteristics

The water utilities responding to this survey include a wide range of plant types and treatment practices.

Utility Age

The age of the utilities was assessed based on the reported year that concrete was first poured in the plant. In the case where a utility reported multiple years (i.e. from upgrades, expansions, or repairs), the oldest date was used for this analysis. Most utilities participating in this survey were built after 1950, although the range varied from one year old to more than 100 years old (Figure 2.1). The average reported year of plant construction was 1967.

Utility Size

The reported capacity of water utilities varied from 0.1 to 655 MGD, with 43% of the utilities producing less than 10 million gallons per day (MGD) (Figure 2.2). Note that some utilities reported total flow from multiple plants, while others reported design flow, rather than actual production.

Coagulant Usage

The most commonly used coagulant chemical used is alum; 70% of participating water utilities currently use alum and nearly 90% report using it at some time at their plant (Figure 2.3). Polyaluminum chloride, ferric chloride, and ferric sulfate are also commonly used. Other coagulants used include polymers, sodium aluminate, and ferrous-based chemicals (see Questions 3 and 4 in Appendix A for a complete list).

Corrosion Inhibitor Usage

Fifty-six percent of water utilities report using a corrosion inhibitor at their plant. The majority of these plants first began using inhibitors in response to the EPA Lead and Copper Rule, and only 15% of utilities have considered using inhibitors to protect treatment plant materials. Most utilities (87%) currently dose the inhibitor after filtration, although 13% add inhibitor either during or after coagulation, which would allow the inhibitor to interact with (and possibly protect) concrete basins.

Utility Concerns About Enhanced Coagulation

More than a quarter of utilities report that they are having to consider changing coagulant type, dose, or coagulation pH to meet the upcoming rules for enhanced coagulation, even though they feel it will worsen aspects of corrosion. It appears that in order to lower pH values to optimize coagulation, most utilities will not be adding supplemental acid or acidified coagulants—only 26% are considering adding acid while 64% report they have no plans to add supplemental acid.

Although the issue of coagulant corrosivity is not reported to be a significant issue, other potential consequences of enhanced coagulation of concern to utilities include:

–Lower pH in the treatment plant and higher dissolved CO_2 in the water (28%)
–Higher chloride or sulfate in the water due to higher doses of coagulants (15%)
–Dissolution of existing protective scale on materials (10%)
–Increased likelihood of copper pitting corrosion in consumer's homes (7%)
–Increased sludge production / handling (3%)
–Cost (2%)
–Other (see Appendix A, Question 7)

As mentioned previously, only 15% of utilities are considering using corrosion inhibitors to protect treatment plant materials from aggressive water. One quarter of utilities have applied epoxy or other protective coatings to protect concrete or metals, and 13% are currently using cathodic protection systems for protection of metals.

Utility Observations of Infrastructure Degradation

Twenty-four percent of all responding utilities report noteworthy degradation of their treatment plant infrastructure due to coagulation, and 22% have had to resurface or repair concrete in their plant. Ten percent have observed signs of rebar corrosion in concrete, and this phenomenon might be more widespread because an additional 20% of utilities said they would not be able to recognize rebar corrosion problems.

Two-thirds of responding utilities are already practicing enhanced coagulation, and 85% of them have not seen any negative effects of their treatment changes. However, some utilities have observed the following detrimental effects:

–Lower pH or higher dissolved CO_2 is causing increased concrete degradation (6%)
–More failure of metal pipes, pumps, or structures in the plant (4%)
–More consumer complaints of red water (2%)
–Increased copper pitting corrosion in consumer's homes (2%).

Concrete poured prior to 1965 is reported to be more durable than newer concrete. Nearly half of utilities reported having concrete from both before and after 1965 in their plant, although most could not tell a difference in performance between the two types. Of the utilities who did see a difference in durability of the two types of concrete, slightly more (7% vs. 5%) found that the old concrete was indeed performing better than the newer concrete.

However, if utilities are sorted according to when they report concrete being poured (141 before 1965 and 111 after 1965), the utilities with older concrete had more problems with infrastructure degradation (Figure 2.4). The most notable difference is that 30% of utilities with concrete poured before 1965 reported having to repair or resurface their concrete, while only 14% of utilities with newer concrete needed such repairs.

Finally, when utilities were asked about money spent repairing or replacing plant infrastructure that had degraded due to coagulation, 68% of utilities reported spending no money on such repairs, while 7% did not know how much was spent. Of the 25% of utilities who did spend money on repairs, the cost ranged from less than $1000 up to $100,000 per year (Figure 2.5), with an average of $13,000 per year. Nineteen plants reported annual costs exceeding $5000 each year. For these plants, if the dollars spent is divided by the average flow rate in MGD, the range of costs was $70–12,600 per MGD per year, with an average of $1850 per MGD per year. Interestingly, only one of these 19 plants experiencing high costs was using ferric chloride or ferric sulfate-- the rest were using alum or PACl.

Finally, it should be noted that under-reporting of costs in the survey is probably quite significant. For example, the City of Manchester depreciates their plant at a rate of $295K per year. In other cases, utilities reported an annual cost attributable to coagulation of $0/year, but they noted that they were building a new treatment plant due to degradation of their existing plant. Clearly, depreciation of assets is not easily or uniformly quantified by most utility personnel, and it is understandably difficult to assign that depreciation directly to coagulant chemicals as was requested in the survey unless a utility had knowledge about such accounting principles.

Miscellaneous Comments on the Survey

An attachment to the survey sent in from Chesterfield County VA highlighted how coagulant feed lines for ferric sulfate had failed twice over a 6 month time period, but replacing the line with a 316 stainless steel injector solved the problem.

Marin County CA noted on their survey that their treatment plant was "falling apart." A follow-up call indicated that they are currently constructing a new plant. While they do not attribute their problems directly to a low coagulation pH (coagulation pH = 6.5 using alum), they do attribute it to a combination of coagulation pH and high chlorine (see Billings case study). Most of the components that are failing are metallic, and in fact, they need a sledgehammer to move many valves. They have found that sacrificial anodes and cathodic protection has worked fairly well in protecting against the chlorine attack.

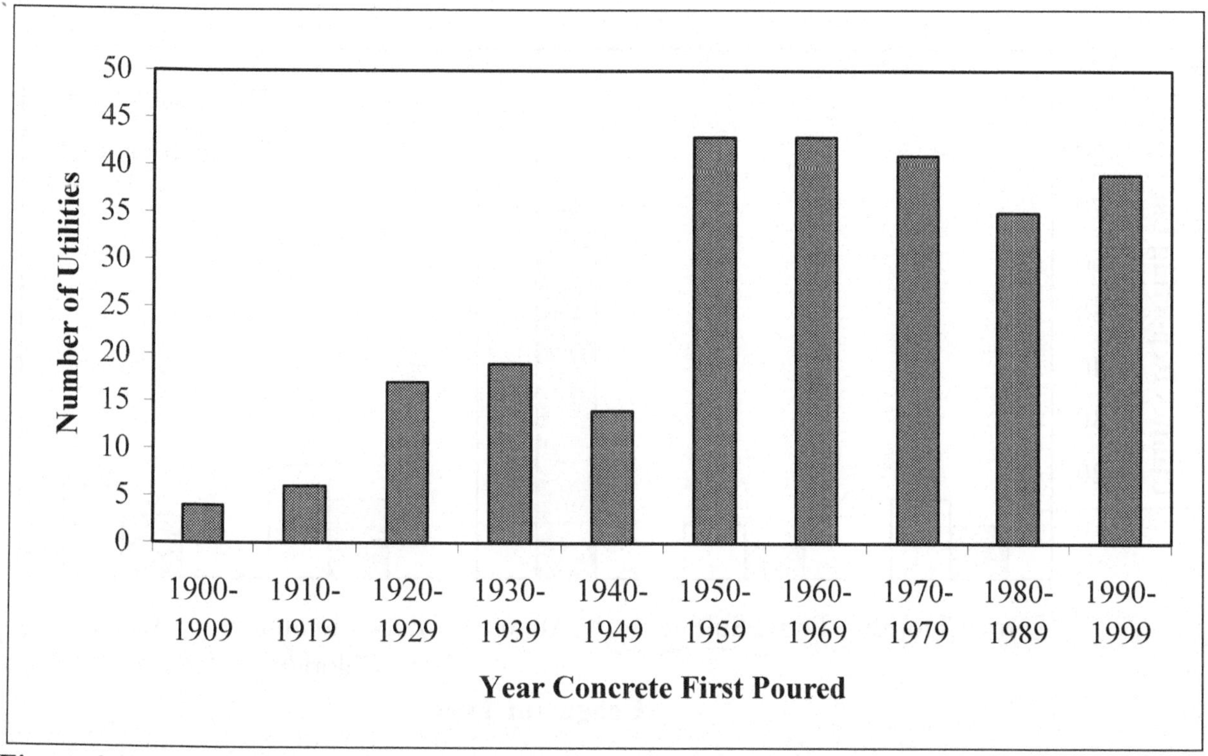

Figure 2.1. Year concrete was first poured at water treatment plants

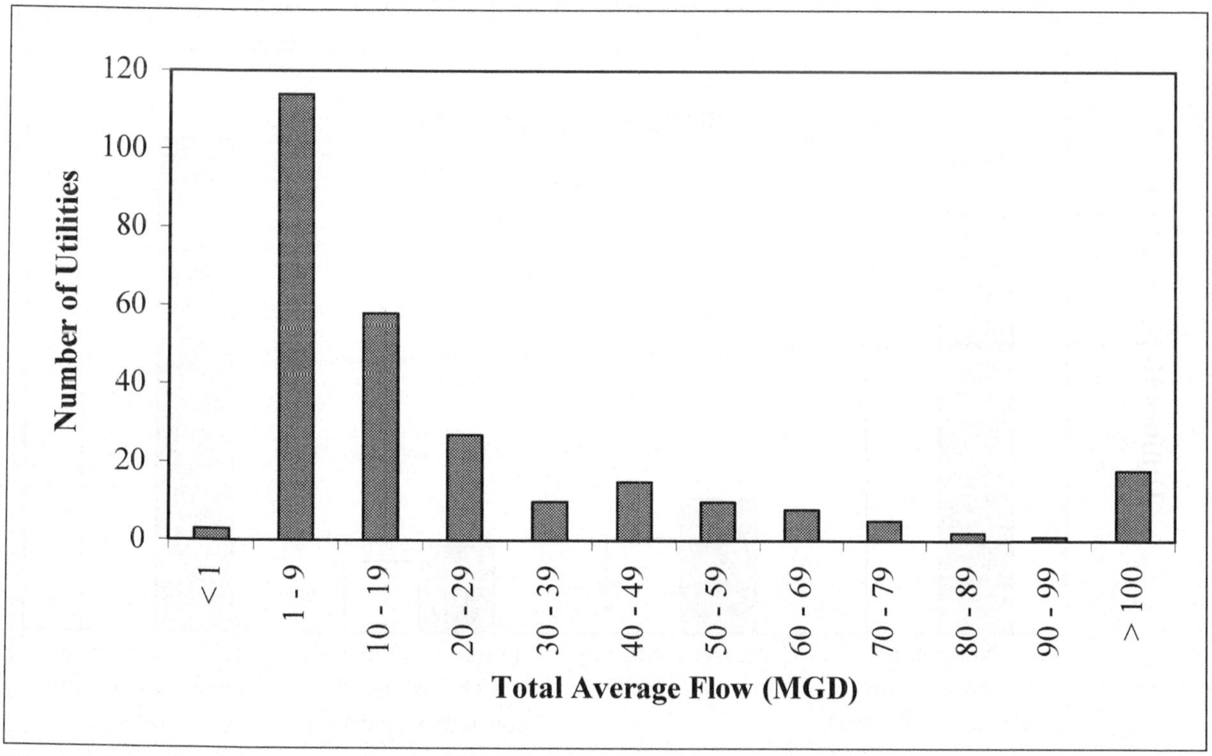

Figure 2.2. Average water production

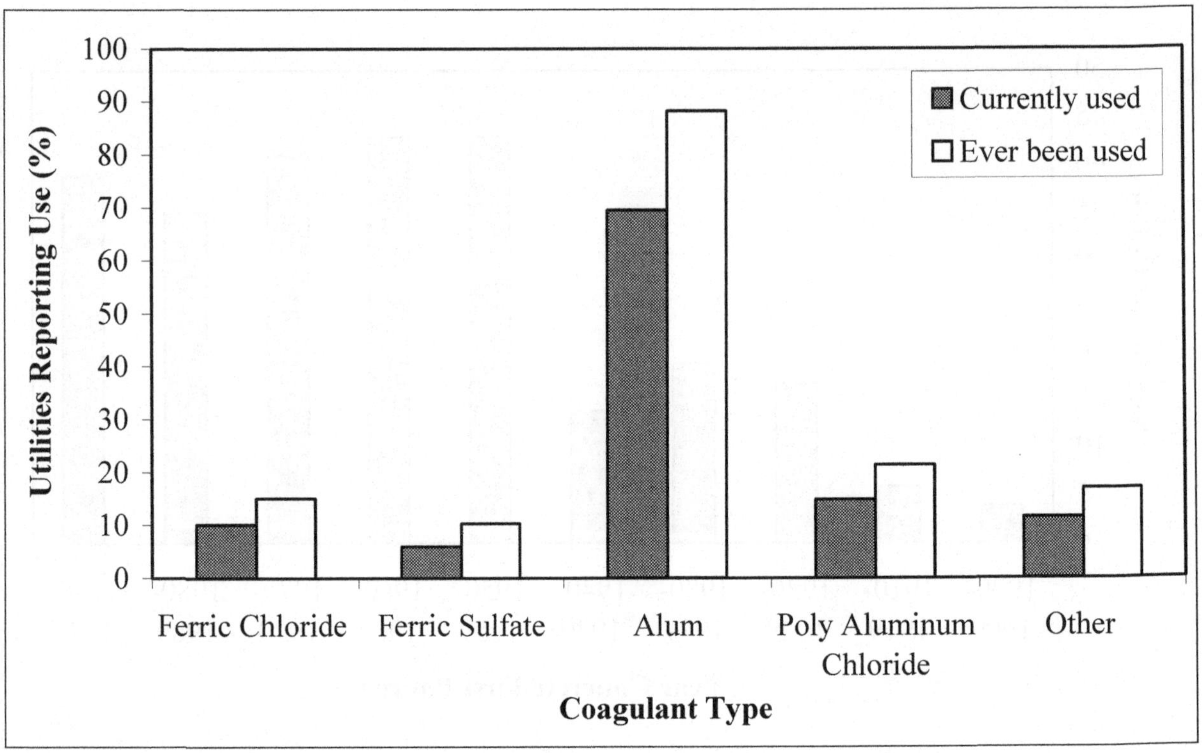

Figure 2.3. Coagulants used at water utilities

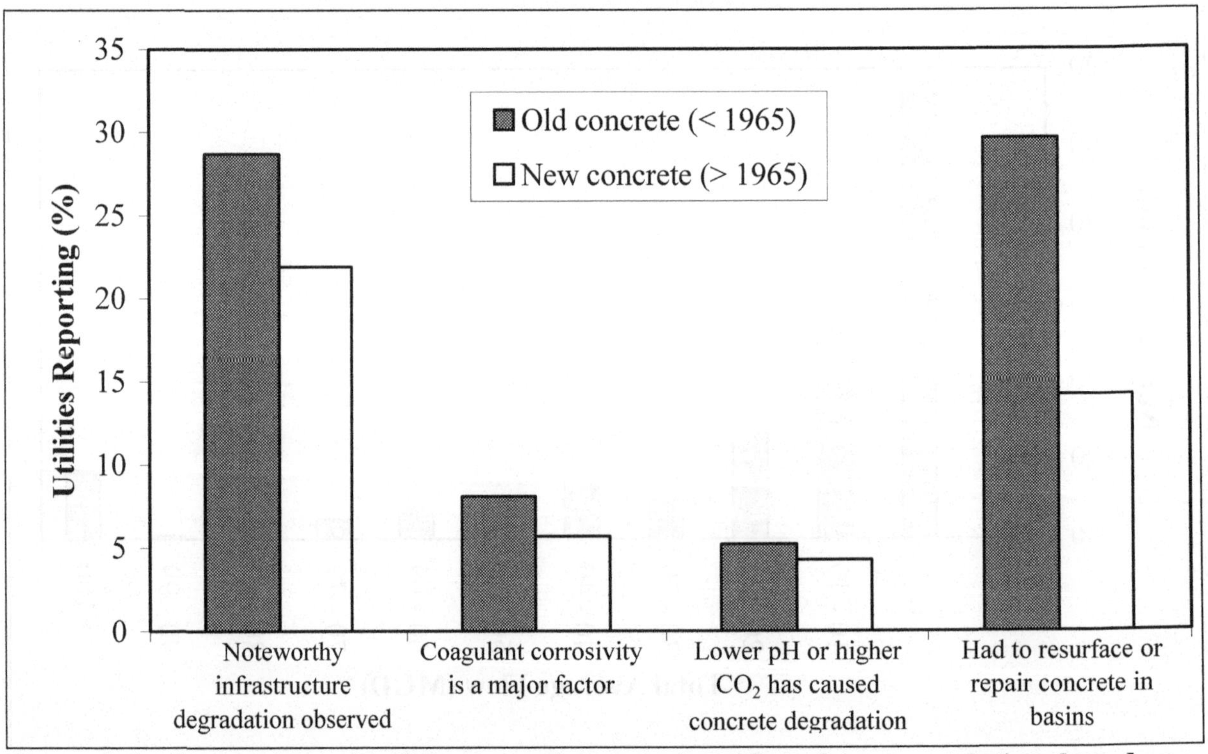

Figure 2.4. Comparison of utilities reporting negative effects from coagulation, based on age of concrete

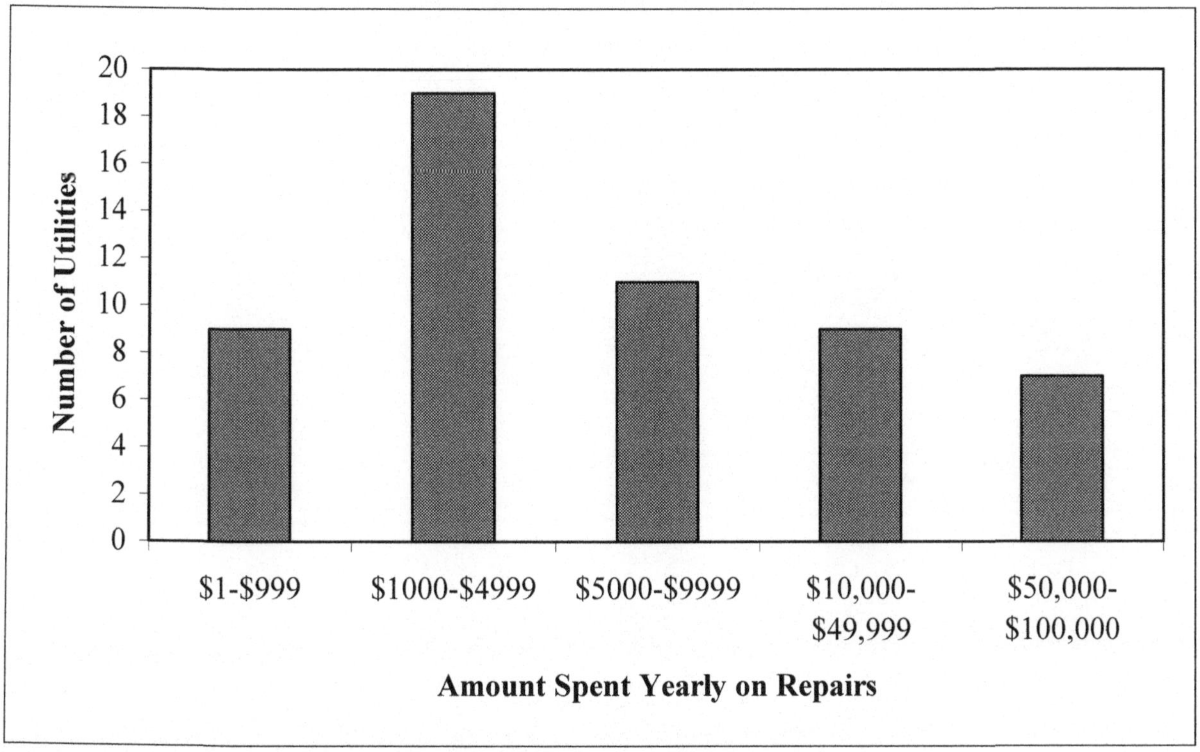

Figure 2.5. Amount spent on repairs due to degradation from corrosion

CHAPTER 3
EXPERIMENTAL TESTING OF INHIBITOR COMPATIBILITY WITH ENHANCED COAGULATION

Although this research project was mainly oriented towards identifying the extent of infrastructure degradation from coagulation and existing practice(s) to prevent it, it was deemed desirable to make two preliminary direct tests of future interest to researchers and utilities. The first test examined whether certain inhibitors were compatible or incompatible with goals of enhanced coagulation. The second tested the relative aggressiveness of three types of coagulants in treating a single source water. The results of the second test are presented in the case studies, whereas results of the first test are presented herein.

It is not commonly realized that enhanced coagulation can dramatically alter the relative desirability of various inhibitors dosed for infrastructure protection. Inhibitors such as polyphosphate, orthophosphate, zinc orthophosphate, zinc, silicates and poly-silicates are most commonly associated with end-of-treatment dosing to protect pipes in water distribution systems. However, some utilities choose to dose inhibitors within the treatment plant to protect plant infrastructure. We currently have an extremely poor understanding of how these inhibitors act within a plant, but it is clear that they are effective under some circumstances.

What are the desirable characteristics of inhibitors dosed to protect drinking water infrastructure at utilities practicing enhanced coagulation? First, the inhibitor must not significantly increase coagulant doses, interfere with NOM removal or interfere with coagulation. After all, NOM removal is the goal of enhanced coagulation, and most corrosion problems arise from the fact that organic matter removal is typically most effective at coagulation pHs near 6. Secondly, dosing of inhibitors is pointless if the inhibitor is removed from the water by the coagulant, since it would not be available to protect plant materials. Thus, the goal of these experiments was to evaluate these aspects of inhibitor compatibility with enhanced coagulation.

MATERIALS AND METHODS

Two experiments evaluated the reactions between coagulant and different inhibitors that might be used to reduce concrete corrosion. All tests were performed on-site at four different treatment plants using their respective source waters. In the first type of test, 2-liter solutions with well formed coagulant flocs were taken directly from the middle of the flocculation basin at the San Diego (CA), Denver (CO) and Lockwood (MT) treatment plants. A control sample was taken by passing the a portion of the water sample through a 0.45 μm pore size filter. Three different inhibitors were then dosed to the water: zinc (1 mg/L as Zn), orthophosphate (0.5 mg/L as P), or silica (15 mg/L as SiO_2). After dosing the inhibitors, the solution was re-adjusted to the original pH using nitric acid or sodium hydroxide. After 15 minutes of mixing, a 0.45 μm filtered sample was collected. The pH of the remaining solution was then incrementally reduced by 0.5 units, mixed 15 minutes, and then filtered again. This procedure was repeated until a solution pH of 5 was achieved. In one test where the treatment plant used a ferric coagulant, a test was conducted in parallel but after dosing alum to the source water at the same dose of metal (mole Fe versus mole Al).

The second type of experiment was conducted using jar tests with raw water from the Buffalo Pound Water Treatment Plant (Regina, Sask.). The jar tests were conducted on-site in Spring and Fall to examine effects of temperature (15 versus 5°C) and different water quality on the results obtained. In each test, three jars were coagulated with different doses of alum, and the other three jars were dosed with the molar equivalent doses of ferric chloride. The alum dose in the first jar was equal to that used at the plant on the day of the test, whereas the other two jars had doses that were 30% higher or 30% lower. The final pH was not adjusted, allowing testing of higher coagulant doses and lower pH as would typically occur during enhanced coagulation. The jars were rapid mixed for 30 seconds at 300 rpm and flocculated at 30 rpm for approximately one hour. After settling, a filtered background sample was taken. The corrosion inhibitors were dosed to the water as noted above, and after mixing for 15 minutes filtered samples were collected.

All samples were analyzed for phosphorus, zinc, silica, iron, aluminum, and arsenic on a JY Ultima Inductive Coupled Plasma-Emission Spectroscopy (ICP-ES) according to Standard Method 3120B using continuous hydride generation.

RESULTS AND DISCUSSION

Experimental results are discussed in two sections. The first examines inhibitor removal by a constant dose of iron or aluminum coagulant but at varying pH. The second examines inhibitor removal when both pH and coagulant doses were changing.

Removal of Inhibitors as a Function of pH

Sorption of cations to the surface of metal hydroxide flocs is expected to increase at higher pH, whereas sorption of anions is expected to decrease. Results of on-site experiments examining inhibitor dosing to waters with coagulants are generally consistent with that expectation (Figure 3.1, Figure 3.2). At San Diego, Denver and Lockwood, between 0 - 60% of a 0.5 mg/L as P dose of ortho-phosphate remained in the water in tests at pH 5 - 8.5, with the rest of the phosphate removed by sorption to the flocs. In some cases, at the lower pHs as little as 5% of the ortho-phosphate remained. In a test at Lockwood, less than 50% of the phosphate was left in the water for iron coagulant when compared to an equivalent dose of alum at the same pH (Figure 3.2). Removal of zinc was highly pH dependent. At pH of 5.5 or less, a maximum of 17% of the Zn^{+2} added to the test waters was removed by coagulation, while at pH 7.0 and 8.5 removal was closer to 70% and 90%, respectively. Less than 10% of all added silica was removed by coagulants regardless of pH.

Removal of Inhibitors when both pH and Coagulant Dose are Changing

On-site tests were conducted at the Buffalo Pound water treatment plant on 9/11/00 and 12/5/00, during which time the water temperature was 15 and 5°C, respectively. The plant was using an alum dose of 82 mg/L, leading to test doses of 63, 82 and 107 mg/L alum in the jar test. These increasing alum doses led to decreasing final pHs of 7.1, 6.9 and 6.5, respectively. The final pHs were the same in tests run for comparison using an equivalent amount of $FeCl_3$ coagulant.

Consistent with observations at the three other plants, removal of phosphate typically increased at higher alum or ferric coagulant dose (Figure 3.3). This is attributable to more floc present for sorption, in addition to the lower coagulation pH at the higher doses. Interestingly, more zinc was removed from the water at higher coagulant doses. Since more floc would tend to increase Zn^{+2} removal while lower pH would tend to decrease Zn^{+2} removal, it appears that the effect of more floc was dominant. As was noted previously, less than 20% of the silica was removed from these waters regardless of coagulant dose and type.

Comparing results on the two different sampling dates, colder temperatures seemed to decrease phosphate removal to flocs, have no effect on silica removal, and increase zinc removal. This must be considered a preliminary result for temperature effects, since other water quality parameters also changed between the two sampling dates, and these factors might also influence inhibitor removal. Overall, the seasonal changes exerted a lesser impact than did coagulation pH.

CONCLUSIONS

Orthophosphate or polyphosphate inhibitors are unlikely to be compatible with goals of enhanced coagulation for protecting treatment plant components, since they are expected to interfere with NOM removal by competition and are easily removed by $Fe(OH)_3$ and $Al(OH)_3$ floc surfaces at lower pHs.

Zn^{+2} inhibitors are more attractive under the lower pH conditions of enhanced coagulation than they are at higher pHs, since only a fraction of the Zn^{+2} is removed at very low pHs. Moreover, cationic materials like Zn^{+2} might actually act like a coagulant and could decrease alum or ferric coagulant doses. Finally, available data do not suggest Zn^{+2} would interfere with NOM removal.

Silica dosing seems increasingly attractive under conditions of enhanced coagulation. Only a small fraction of silica is removed by the coagulant at any pH (Figure 3.2), and silica interference with NOM removal is actually eliminated as the pH is depressed to 6.0 or lower. However, silica interference with NOM removal is very significant at higher pHs.

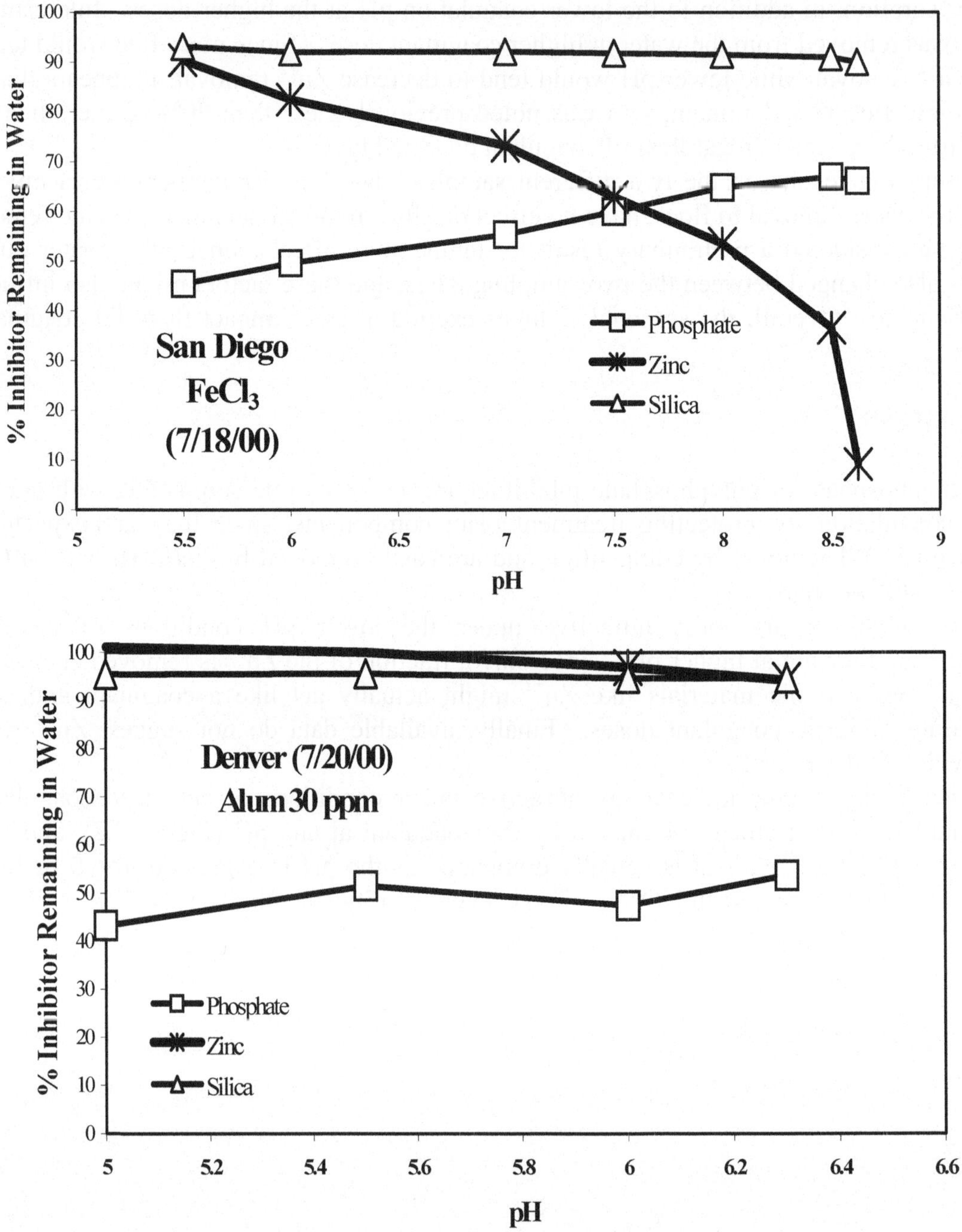

Figure 3.1. Percentage of inhibitor doses remaining in water as a function of pH for on-site tests at Denver and San Diego

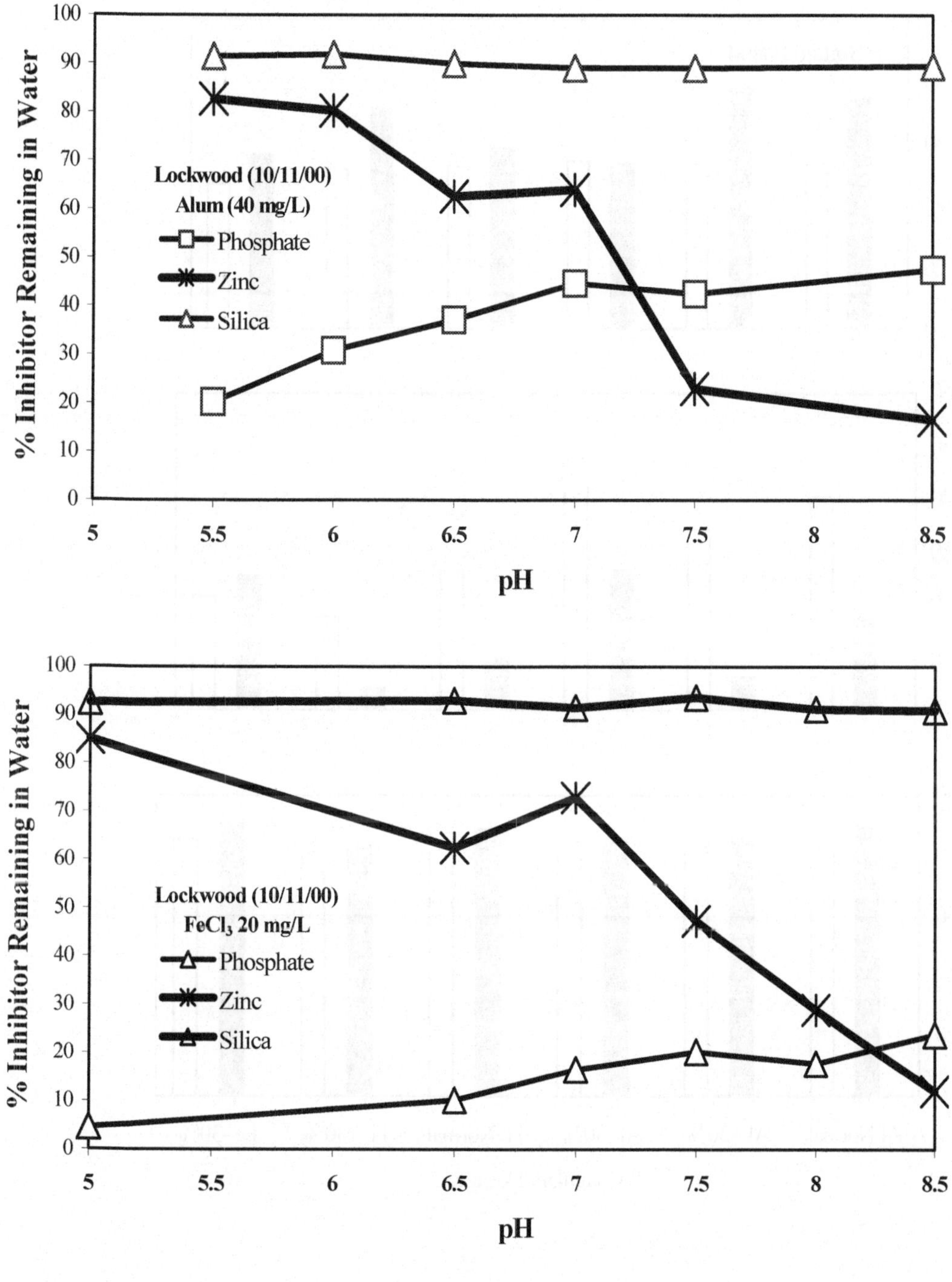

Figure 3.2. Percentage of inhibitor remaining in the water after alum and ferric coagulation at Lockwood

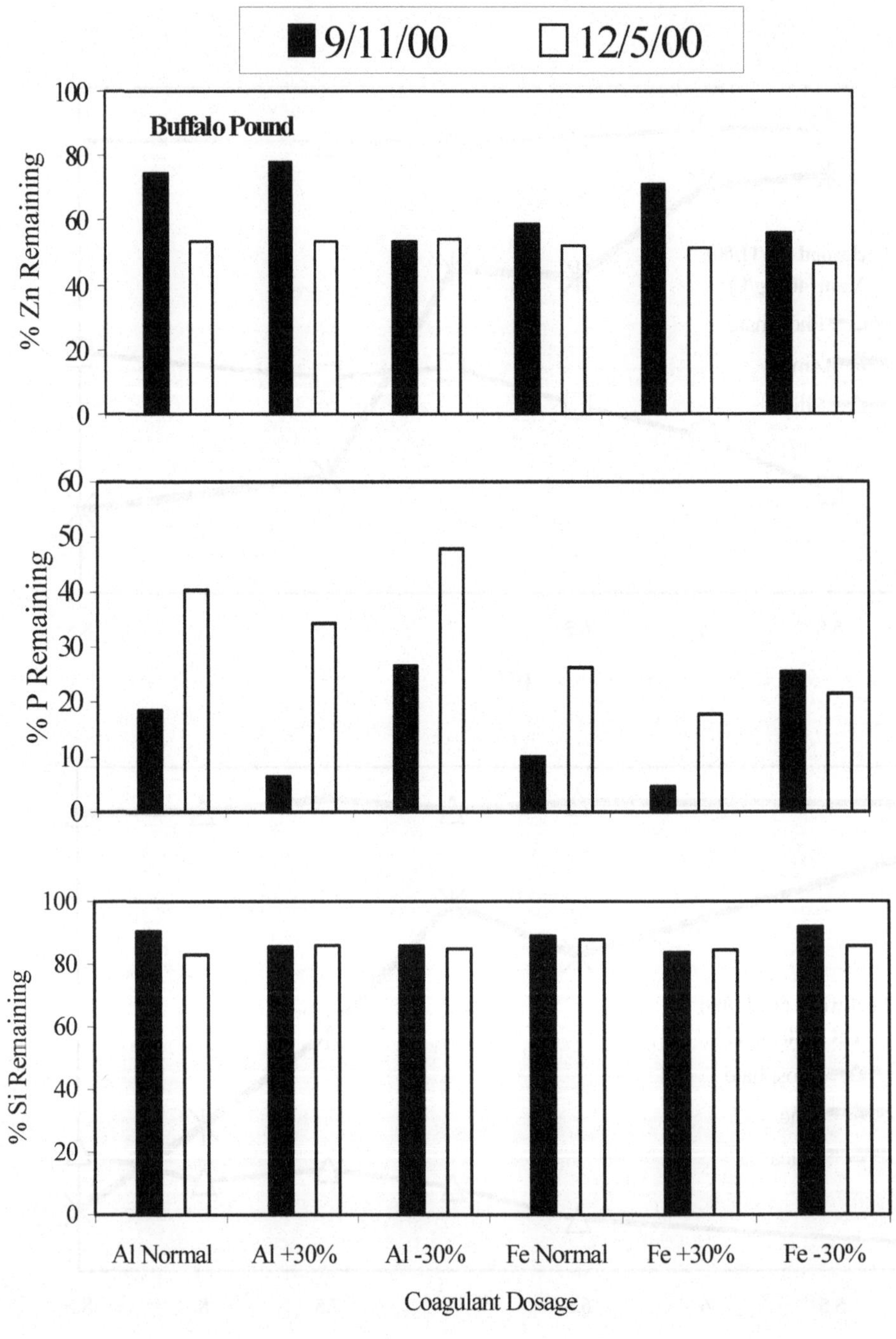

Figure 3.3. Percent of inhibitor remaining in water after dosing to Buffalo Pound water on two different dates

APPENDIX A
RAW DATA FROM UTILITY SURVEY

The survey questions and possible answers appear as presented to each utility personnel. On the far left in parenthesis, the top number indicates the total number of respondents to each question, and the number of responses for each answer is listed in parenthesis to the far left of each possible answer. This is also presented in terms of percentage responses to each question, presented at the immediate left of each answer.

1. Our records indicate that you operate at least one surface water treatment plant and are practicing coagulation. Is this correct?

(275)
(271) 99% A. Yes
(4) 1% B. No

2. Has there been any noteworthy degradation of your plant infrastructure (i.e., concrete, pipes, pumps, metal structures) that you believe is due, at least in part, to coagulation chemicals?

(264)
(64) 24% A. Yes
(197) 75% B. No
(3) 1% not sure

3. Which of the following coagulants have ever been used in your treatment plant? (please circle all that apply)

(271)

(41)	15%	A. Ferric Chloride		
(28)	10%	B. Ferric Sulfate		
(239)	88%	C. Alum or Aluminum Sulfate		
(58)	21%	D. Poly Aluminum Chloride		
(46)	17%	E. Other		
		(15)	5.5%	Cationic polymer
		(11)	4.1%	Polymer
		(5)	1.8%	Lime
		(4)	1.5%	Aluminum chlorahydrate
		(3)	1.1%	Sodium aluminate
		(3)	1.1%	Ferrous sulfate
		(2)	0.7%	Anionic polymer
		(1)	0.4%	Ferrous chloride

4. Which coagulant(s) do you currently use during the year? (if you use more than one type, e.g. alum in the summer and ferric chloride in the winter, please circle all that apply)

(269)

(27)	10%	A. Ferric Chloride		
(16)	6%	B. Ferric Sulfate		
(187)	70%	C. Alum or Aluminum Sulfate		
(40)	15%	D. Poly Aluminum Chloride		
(31)	12%	E. Other		
		(15)	5.6%	Cationic polymer
		(10)	3.7%	Polymer
		(4)	1.5%	Aluminum chlorahydrate
		(2)	0.7%	Lime
		(1)	0.4%	Each: Ferrous chloride; Ferrous sulfate; Anionic polymer; Poly aluminum hydroychloro sulfate; Sodium aluminate

5. Has the issue of corrosion from the coagulants been a major factor in coagulant use?

(271)

(17)	6%	A. Yes
(254)	94%	B. No

6. To meet forthcoming regulations for enhanced coagulation, are you being forced to consider changing coagulant type, dose or coagulation pH, even though you think it will worsen aspects of corrosion?

(262)

(71)	27%	A. Yes
(188)	72%	B. No
(3)	1%	possibly

7. If you are making (or have made) a change to coagulation to improve removal of organic matter, what consequences of the change most concern you? (please circle all that apply)

(227)

(64) 28% A. Lower pH in the treatment plant and higher dissolved CO_2 in the water

(33) 15% B. Higher chloride or sulfate in the water due to higher doses of coagulants

(23) 10% C. Dissolution of existing protective scale on materials

(15) 7% D. Increased likelihood of copper pitting corrosion in consumer's homes

(23) 10% E. Other

- (8) 3% increased sludge production / handling
- (6) 2% cost
- (1) 0.5% Each: Al levels; turbidity; LCR compliance; Phosphate pptn; $FeSO_4$ corrosivity; staining; iron; reduced alkalinity; application monitoring; meeting particle removal and TOC removal requirements at same time; Mn control across filters (at lower pH, absorptive capacity of media is less)

(133) 59% F. None of the above are a concern

8. If you have already made a change to coagulation in order to improve removal of organic matter, have you seen evidence of the following? (please circle all that apply)

(179)

(10) 6% A. Lower pH or higher dissolved CO_2 is causing increased concrete degradation

(7) 4% B. More failure of metal pipes, pumps, or structures in the plant

(4) 2% C. More consumer complaints of red water

(3) 2% D. Increased copper pitting corrosion in consumer's homes

(2) 1% E. Other corrosion problems (phosphate pptn; corrosion of plant structures)

(153) 85% F. None of the above are a concern

9. Which of the following statements best represents your current view regarding addition of supplemental acid (i.e., HCl, H_2SO_4) or coagulant with extra acid added to it (i.e., acidified alum) to reduce the pH of coagulation, improve removal of organic matter, reduce costs of treatment, or improve coagulation performance?

(261)

(166) 64% A. We have never added acid and have no plans to do so in the future

(68) 26% B. We have not added acid in the past but we might consider it in the future

(21) 8% C. We are currently adding acid and have no plans to change

(7) 3% D. We have added acid in the past, but we no longer do so

10. Does your utility currently use, or have you ever used, corrosion inhibitors (other than pH or alkalinity adjustment)?

(269)

(150) 56% A. Yes

(119) 44% B. No

11. Where is this inhibitor dosed? (circle the answer that best applies)
(145)
(0) 0% A. Before coagulant is added
(1) 1% B. At the same time as the coagulant
(17) 12% C. After coagulant addition but before filtration
(126) 87% D. After filtration
(2) 1% other

16. Did your utility begin dosing the corrosion inhibitors mostly to improve control of lead and copper corrosion (i.e., because of the EPA Lead and Copper Rule)?
(147)
(83) 56% A. Yes
(64) 44% B. No

18. Have you considered dosing inhibitors to protect your treatment plant materials?
(265)
(39) 15% A. Yes
(226) 85% B. No

19. Have you applied epoxy or other protective coatings to protect your concrete and metals from aggressive water produced during coagulation?
(267)
(68) 25% A. Yes
(200) 75% B. No

20. Have you had to resurface or repair concrete in your coagulation basins or filters?
(266)
(58) 22% A. Yes
(200) 75% B. No
(9) 3% Yes, but not due to coagulant corrosivity

21. Are you using anodic or cathodic protection systems for exposed metal in the coagulation basins?
(269)
(34) 13% A. Yes
(235) 87% B. No

22. Are there signs of rebar corrosion in your concrete?
(267)
(26) 10% A. Yes
(180) 67% B. No
(53) 20% C. Would not be able to recognize this problem
(8) 3% Yes, but not due to chemicals

23. Please indicate the approximate year when concrete in the main part of your plant was first poured.

(265) See Figure 2.1

24. We define "new" concrete as that poured after 1965, and "old" concrete as that poured before 1965. In terms of cracking or etching of the concrete surfaces that are in contact with water, which of the following describes your case?

(262)

(144)	55%	A. We only have one type of concrete
(19)	7%	B. We have both types of concrete; the old concrete is holding up better than new
(14)	5%	C. We have both types of concrete; the old concrete is not better than the new
(85)	32%	D. We have both types of concrete, and I cannot tell the difference

25. About how much do you spend each year caring for your concrete, replacing your concrete, or replacing pipes, pumps and other materials where degradation is believed to result all or in part from coagulation?

(217) See Figure 2.5

26. What is the total average flow at your treatment plant?

(271) See Figure 2.2

REFERENCES

Anonymous. 1983. A Spec Odyssey--Sulfate Resistance for the 1980's. *Journal of the American Concrete Institute*, 80(6):551-552.

Anonymous. 1991. Some Fly Ashes Reduce Sulfate Resistance. *Aberdeen's Concrete Construction*, 36(2):225-226.

Anonymous. 1992. Corrosion-Resistant System Protects Concrete. *Water/Engineering & Management*, 139(10):13-14.

Anonymous. 1994. New Coating Protects Water Tank Linings. *Public Works*, 125(2):38-39.

Anonymous. 1995. Urethane Protects Precipitator. *Journal of Protective Coatings and Linings*, 12(1):46-47.

Anonymous. 1997a. Protecting Water Storage Tanks in an Era of Environmental Compliance. *Public Works*, 128(5):34-37.

Anonymous. 1997b. Cutting Off Corrosion Damage. *Water/Engineering & Management*, 144(4):10.

Al-Amoudi, O.S.B., M. Maslehuddin and M.M. Saadi. 1995. Effect of Magnesium Sulfate and Sodium Sulfate on the Durability Performance of Plain and Blended Cements. *ACI Materials Journal*, 92(1):15-24.

Alfonsson, E., A. Iverson and J. Olsson. 1999. Stainless Steels in Sewage Treatment Plants. In *Proc. of the 14th International Corrosion Congress*. Cape Town, South Africa: International Corrosion Council.

Al-Gahtani, A.S., M. Ibrahim, M. Maslehuddin and A.A. Almusallam. 1999. Performance of Concrete Surface Treatment Systems. *Concrete International*, 21(1):64-68.

Alexander, M., B. Addis, and J. Basson. 1994. Case Studies Using a Novel Method to Assess Aggressiveness of Waters to Concrete. *ACI Materials Journal*, 91(2):188-196.

Alhozaimy, A., P. Soroushian and F. Mirza. 1996. Effects of Curing Conditions and Age On Chloride Permeability of Fly Ash Mortar. *ACI Materials Journal*, 93(1):87-95.

Alonso, C., C. Andrade, M. Izquierdo, X.R. Novoa and M.C. Perez. 1998. Effect of Protective Oxide Scales in the Macrogalvanic Behaviour of Concrete Reinforcements. *Corrosion Science*, 40(8):1379-1389.

AWWA. 2001. Dawn of the Replacement Era: Reinvesting in Drinking Water Infrastructure. AWWA Government Affairs Office, Washington DC.

AwwaRF (Awwa Research Foundation) and DVGW-Technologiezentrum Wasser. 1996. *Internal Corrosion of Water Distribution Systems, 2nd Edition.* Denver, CO: AwwaRF.

Baweja, D., R. Roper and V. Sirivivatnanon. 1998. Chloride-Induced Steel Corrosion in Concrete: Part 1--Corrosion Rates, Corrosion Activity, and Attack Areas. *ACI Materials Journal*, 95(3):207-217.

Baweja, D., H. Roper and V. Sirivivatnanon. 1999. Chloride-Induced Steel Corrosion in Concrete: Part 2--Gravimetric and Electrochemical Comparisons. *ACI Materials Journal*, 96(3):306-313.

Baylis, J.R. 1927. Treatment of Water to Prevent Corrosion. *Industrial and Engineering Chemistry*, 19(7):777-781.

Benedict, R.L. 1990. Corrosion Protection of Concrete Cylinder Pipe. *Materials Performance*, 29(2):22-29.

Benedict, R.L., J.G. Ott II, D.H. Marshall and D. White. 1997. Cathodic Protection of Prestressed Concrete Cylinder Pipe Utilizing Zinc Anodes. *Materials Performance*, 36(5):12-17.

Berke, N.S. and M.C. Hicks. 1994. Predicting Chloride Profiles in Concrete. *Corrosion*, 50(3):234-239.

Bierwagen, G.P. 1998. Corrosion and Its Control by Coatings. In *Organic Coatings for Corrosion Control.* Washington, DC: American Chemical Society.

Billings, C.H. 1996. NOM: Copper Corrosion. *Public Works*, 127(11):74-76.

Bishara, S.W. 1991. Rapid, Accurate Method for Determining Water-Soluble Chloride in Concrete, Cement, Mortar, and Aggregate: Application to Quantitative Study of Chloride Ion Distribution in Aged Concrete. *ACI Materials Journal*, 88(3):265-270.

Bromley, D. 1995. Solvent-Free Epoxy Coatings for Potable Water. *Journal of the Indian Water Works Association*, 27(3):163-165.

Broomfield, J.P. 2000. Carbonation and Its Effects in Reinforced Concrete. *Materials Performance*, 39(1):64-67.

Campbell, H.S. and M.E.D. Turner. 1983. The Influence of Trace Organics on Scale Formation and Corrosion. *Journal of the Institute for Water Engineering and Science*, 4:55-72.

Cannon, R.W. 1999. Sulfate Resistance of Roller Compacted Concrete, Discussion. *ACI Materials Journal*, 96(3):410-411.

Carr, F.H. 1991. Salt Water Exhibit Tanks Growing Old Prematurely. *ENR*, 227(9):16.

Carson, H.Y. 1927. Cement-Lined Water Mains. *Industrial and Engineering Chemistry*, 19(7):781-783.

Castellote, M., C. Andrade and M.C. Alonso. 1999. Changes in Concrete Pore Size Distribution Due to Electrochemical Chloride Migration Trials. *ACI Materials Journal*, 96(3):314-319.

Chappell, E.L. 1930. Chemical Characteristics of Cement Pipe Lining. *Industrial and Engineering Chemistry*, 22(11):1203-1206.

Chmelar, P. 1992. Durable Coatings Protect Concrete Structures. *Chemical Engineering*, 99(4):171-172.

Cohen, M.D. and A. Bentur. 1988. Durability of Portland Cement-Silica Fume Pastes in Magnesium Sulfate and Sodium Sulfate Solutions. *ACI Materials Journal*, 85(3):148-157.

Cohen, M.D. and B. Mather. 1991. Sulfate Attack on Concrete--Research Needs. *ACI Materials Journal*, 88(1):62-69.

Collins, W.D., R.E. Weyers and I.L. Al-Qadi. 1993. Chemical Treatment of Corroding Steel Reinforcement After Removal of Chloride-Contaminated Concrete. *Corrosion*, 49(1):74-88.

Daczko, J.A., D.A. Johnson and S.L. Amey. 1997. Decreasing Concrete Sewer Pipe Degradation Using Admixtures. *Materials Performance*, 36(1):51-56.

Davis, J. 2000. What You Can't See...Can Hurt You. *Materials Performance*, 39(5):44-46.

Day, R.W. 1995. Damage to Concrete Flatwork From Sulfate Attack. *Journal of Performance of Constructed Facilities*, 9(11):302-310.

Detwiler, R.J., C.A. Fapohunda and J. Natale. 1994. Use of Supplementary Cementing Materials to Increase the Resistance to Chloride Ion Penetration of Concretes Cured at Elevated Temperatures. *ACI Materials Journal*, 91(1):63-66.

Detwiler, R.J., K.O. Kjellsen and O.E. Gjorv. 1991. Resistance to Chloride Intrusion of Concrete Cured at Different Temperatures. *ACI Materials Journal*, 88(1):19-24.

Detwiler, R.J., D.A. Whiting and E.S. Lagergren. 1999. Statistical Approach to Ingress of Chloride Ions in Silica Fume Concrete for Bridge Decks. *ACI Materials Journal*, 96(6):670-675.

Dillon, C.P. 1994. Behavior of Lead in Organic Sulfate Environments. *Materials Performance*, 33(12):54.

Dodrill, D. and M. Edwards. 1995. Corrosion Control on the Basis of Utility Experience. *Jour. AWWA*, 87(7):74-85.

Edwards, M. 1997. Predicting DOC Removal During Enhanced Coagulation. *Jour. AWWA*, 89(5):78-89.

Edwards, M., S. Reiber and J.F. Ferguson. 1994. On Pitting Corrosion of Copper. *Jour. AWWA*, 86(7):74-90.

Edwards, M., M.R. Schock and T.E. Meyer. 1996. pH, Alkalinity and Copper Corrosion By-Products. *Jour. AWWA*, 88(3):81-94.

Elfstrom Broo, A., B. Berghult and T. Hedberg. 1998. Copper Corrosion in Water Distribution Systems--The Influence of Natural Organic Matter (NOM) on the Solubility of Copper Corrosion Products. *Corrosion Science*, 40(9):1479-1489.

Ellis, D.J. and G.C. Moore. 1993. Materials and Coating Selection for Water Supply and Sewerage, an Essential Component of Asset Management. *Corrosion Australasia*, 18(4):11-14.

Figg, J. 1983. Chloride and Sulphate Attack on Concrete. *Chemistry and Industry*, 20:770-775.

Finch, D. 1995. Coating Secondary Containment Structures. *Plant Engineering*, 49:70-74.

Flente, M.E. and R.J. Sweitzer. 1955. Solution Effects of Water on Cement and Concrete in Pipe. *Jour. AWWA*, 47(12):1173-1194.

Foscante, R.E. and H.H. Kline. 1988. Coating Concrete--An Overview. *Materials Performance*, 27(9):34-36.

Freeman, R.B. and R.L. Carrasquillo. 1995. Adjustments in Gypsum Content for Production of Sulfate-Resistant Blended Cements Containing High-Calcium Fly Ash. *ACI Materials Journal*, 92(4):411-418.

Ghafoori, N. and Z. Zhang. 1998. Sulfate Resistance of Roller Compacted Concrete. *ACI Materials Journal*, 95(4):347-355.

Giacasso, P. 1988. Corrosion du beton dans une station de traitement d'eau de boisson d'eau desequilibre calco-carbonique. Journees Italo-Franco-Suisses AGHTM, SVGW, Ferdergasacqua. Avignon, Mai.

Glass, G.K. and N.R. Buenfeld. 2000. The Inhibitive Effects of Electrochemical Treatment Applied to Steel in Concrete. *Corrosion Science*, 42(6):923-927.

Glass, G.K., B. Reddy and N.R. Buenfeld. 2000. Corrosion Inhibition in Concrete Arising from its Acid Neutralisation Capacity. *Corrosion Science*, 42(9):1587-1598.

Glass, G.K., J.-Z. Zhang and N.R. Buenfeld. 1995. Chloride Ion Barrier Properties of Small Electric Fields in the Protection of Steel in Concrete. *Corrosion*, 51(9):721-726.

Gotoh, K. and T. Shirozu. 1997. Relining Techniques for Pipes and Concrete Structures. *Water Supply*, 15(1):75-82.

Grabowski, E., B. Czarnecki, J.E. Gillott, C.R. Duggan and J.F. Scott. 1992. Rapid Test of Concrete Expansivity Due to Internal Sulfate Attack. *ACI Materials Journal*, 89(5):469-480.

Greenfield, T.K. 1994. Specifying Coatings for Concrete Surfaces. *Materials Performance*, 33(7):37-41.

Groves, B. 1999. Concrete and Masonry Refurbishment in the Water Industry. *Concrete*, 33(2):10-11.

Gu, P., J.J. Beaudoin, M.-H. Zhang and V.M. Malhotra. 1999. Performance of Steel Reinforcement in Portland Cement and High-Volume Fly Ash Concretes Exposed to Chloride Solution. *ACI Materials Journal*, 96(5):551-558.

Gu, P., J.J. Beaudoin, M.-H. Zhang and V.M. Malhotra. 2000. Performance of Reinforcing Steel in Concrete Containing Silica Fume and Blast-Furnace Slag Ponded with Sodium Chloride Solution. *ACI Materials Journal*, 97(3):254-262.

Gu, P., S. Elliott, R. Hristova, J.J. Beaudoin, R. Brousseau and B. Baldock. 1997. A Study of Corrosion Inhibitor Performance in Chloride Contaminated Concrete by Electrochemical Impedance Spectroscopy. *ACI Materials Journal*, 94(5):385-395.

Gummow, R.A. 2000. CP Corrosion Control of Municipal Infrastructure. *Materials Performance*, 39(2):26-32.

Haque, M.N. and M. Kawamura. 1992. Carbonation and Chloride-Induced Corrosion of Reinforcement in Fly Ash Concretes. *ACI Materials Journal*, 89(1):41-48.

Haque, M.N. and O.A. Kayyali. 1995. Aspects of Chloride Ion Determination in Concrete. *ACI Materials Journal*, 92(5):532-541.

Haque, M.N., O.A. Kayyali and M.K. Gopalan. 1992. Fly Ash Reduces Harmful Chloride Ions in Concrete. *ACI Materials Journal*, 89(3):238-241.

Harper, W.B. 1991. Environmental and Health Concerns Relating to Water and Wastewater Coating and Painting Projects. *Materials Performance*, 30(2):56-61.

Hime, W.G. 1994. Chloride-Caused Corrosion of Steel in Concrete: A New Historical Perspective. *Concrete International*, 16(5):56-61.

Hooton, R.D. 1993. Influence of Silica Fume Replacement of Cement on Physical Properties and Resistance to Sulfate Attack, Freezing and Thawing, and Alkali-Silica Reactivity. *ACI Materials Journal*, 90(2):143-151.

Hope, B.D. and A.K.C. Ip. 1987. Chloride Corrosion Threshold in Concrete. *ACI Materials Journal*, 84(4):306-314.

Hou, J. and D.D.L. Chung. 2000. Effect of Admixtures in Concrete on the Corrosion Resistance of Steel Reinforced Concrete. *Corrosion Science*, 42(9):1489-1507.

Ilaria, J.E. 1995. Prolong the Life of Concrete. *Chemical Engineering Progress*, 91(7):67-71.

Janotka, I. and L. Stevula. 1998. Effect of Bentonite and Zeolite on Durability of Cement Suspension Under Sulfate Attack. *ACI Materials Journal*, 95(6):710-715.

Johnson, J.B. and B.S. Skerry. 1998. An Aspect of Concrete Protection by Surface Coating. In Proc. *Organic Coatings for Corrosion Control.* Washington, DC: American Chemical Society.

Johnston, C.D. 1994. Deicer Salt Scaling Resistance and Chloride Permeability. *Concrete International*, 16(8):48-55.

Khan, M.S. 1997. Chloride Ion Analysis by AASHTO Method. *Concrete International*, 19(10):67-69.

Kirmeyer, G.J., G. Pierson, J. Clement, A. Sandvig, V. Snoeyink, W. Kriven and A. Camper. 2000. *Distribution System Water Quality Changes Following Corrosion Control Strategies*. Denver, CO: AwwaRF and AWWA.

Kline, H.H. 1983. Coatings and Linings in Waste Water Treatment Facilities. In *Proc. of the NACE International Corrosion Forum*. Anaheim, CA: National Association of Corrosion Engineers.

Korshin, G.V., J.F. Ferguson and A.N. Lancaster. 2000. Influence of Natural Organic Matter on the Corrosion of Leaded Brass in Potable Water. *Corrosion Science*, 42(1):53-66.

Korshin, G.V., S.A.L. Perry and J.F. Ferguson. 1996. Influence of NOM on Copper Corrosion. *Journal AWWA*, 88(7):36-47.

Kosmatka, S.H. and W.C. Panarese. 1992. *Design and Control of Concrete Mixtures. 13th Edition.* Washington DC: Portland Cement Association.

Krajci, L. and I. Janotka. 2000. Measurement Techniques for Rapid Assessment of Carbonation in Concrete. *ACI Materials Journal*, 97(2):168-171.

Kurtis, K.E., P.J.M. Monteiro and S.M. Madanat. 2000. Empirical Models to Predict Concrete Expansion Caused by Sulfate Attack. *ACI Materials Journal*, 97(2):156-161.

Kvech, S.J. and M. Edwards. 2001. The Role of Aluminosilicate Deposits in Lead and Copper Corrosion. *Jour. AWWA*, 93(11), 104-112.

Larson, T.E. 1966. Chemical Control of Corrosion. *Jour. AWWA*, 58(3):354-362.

Lauer, W.C., S.R. Lohman and M.D. Rose. 1999. Corrosive Water Can Protect Pipes. Denver, CO: Denver Water.

Legrand, L. and P. Leroy. 1990. *Prevention of Corrosion and Scaling in Water Supply Systems.* West Sussex, England: Ellis Horwood Limited.

Li, Z., J. Peng and B. Ma. 1999. Investigation of Chloride Diffusion for High-Performance Concrete Containing Fly Ash, Microsilica, and Chemical Admixtures. *ACI Materials Journal*, 96(3):391-396.

Loto, C.A. 1992. Effect of Inhibitors and Admixed Chloride on Electrochemical Corrosion Behavior of Mild Steel Reinforcement in Concrete in Seawater. *Corrosion*, 48(9):759-763.

Loulizi, A., I.L. Al-Qadi and B.K. Diefenderfer. 2000. Effects of Nitrite-Based Corrosion Inhibitor on Concrete's Rapid Chloride Permeability Values and its Dielectric Properties. *ACI Materials Journal*, 97(4):465-471.

Luma, C. 1998. Protect Concrete from Corrosion. *Chemical Engineering*, 105(12):149-150.

Luping, T. and L.-O. Nilsson. 1992. Rapid Determination of the Chloride Diffusivity in Concrete By Applying an Electrical Field. *ACI Materials Journal*, 89(1):49-53.

Mailvaganam, N.P. 1992. *Repair and Protection of Concrete Structures.* Boca Raton, FL: CRC Press, Inc.

Malhotra, V.M., M.H. Zhang and G.H. Leaman. 2000. Long-Term Performance of Steel Reinforcing Bars in Portland Cement Concrete and Concrete Incorporating Moderate and High Volumes of ASTM Class F Fly Ash. *ACI Materials Journal*, 97(4):409-417.

Mangat, P.S. and J.M. Khatib. 1995. Influence of Fly Ash, Silica Fume, and Slag on Sulfate Resistance of Concrete. *ACI Materials Journal*, 92(5):542-552.

McGovern, M.S. 1999. Can Coatings Protect Wastewater Treatment Systems? *Aberdeen's Concrete Construction*, 44(4):53.

McGovern, M.S. 2000. A Clear View of Sealers. *Concrete Construction*, 45(1):53,55,57-58.

Mehta, P.K. 1986. Effect of Fly Ash Composition on Sulfate Resistance of Cement. *Journal of the American Concrete Institute*, 83(6):994-1000.

Mehta, P.K. 2000. Sulfate Attack on Concrete: Separating Myths From Reality. *Concrete International*, 22(8):57-61.

Montemor, M.F., A.M.P. Simoes and M.G.S. Ferreira. 1998. Analytical Characterization of the Passive Film Formed on Steel in Solutions Simulating the Concrete Interstitial Electrolyte. *Corrosion*, 54(5):347-353.

Morris, W., M. Vazquez and S.R. De Sanchez. 2000. Efficiency of Coatings Applied on Rebars in Concrete. *Journal of Materials Science*, 35(8):1885-1890.

Munger, C.G. 1984. *Corrosion Prevention by Protective Coatings.* Houston, TX: National Association of Corrosion Engineers.

Munger, C.G. 1990. Coating Concrete. *Materials Performance*, 29(12):41-42.

Nagataki, S., N. Otsuki, T.-H. Wee and K. Nakashita. 1993. Condensation of Chloride Ion in Hardened Cement Matrix Materials and on Embedded Steel Bars. *ACI Materials Journal*, 90(4):323-332.

Nicholas, D. and G. Moore. 1999. Materials Engineering in Water and Wastewater Treatment Plant Design. In *Proc. of the Australian Water Works Association Conference*. Adelaide, Australia: AWWA.

Nmai, C.K., S.A. Farrington and G.S. Bobrowski. 1992. Organic-Based Corrosion-Inhibiting Admixture for Reinforced Concrete. *Concrete International*, 14(4):45-51.

Paul, J.H. 1998. Extending the Life of Concrete Repairs. *Concrete International*, 20(3):62-66.

Pey, J.-L. 1997. Corrosion Protection of Pipes, Fittings, and Component Pieces of Water Treatment and Pumping Stations. *Anti-Corrosion Methods and Materials*, 44(2):94-99.

Pyc, W.A., R.E. Weyers, M.M. Sprinkel, R.M. Weyers, D.W. Mokarem and J.G. Dillard. 2000. Performance of Epoxy-Coated Reinforcing Steel. *Concrete International*, 22(2):57-62.

Rasheeduzzafar, F.H. Dakhil, A.S. Al-Gahtani, S.S. Al-Saadoun and M.A. Bader. 1990. Influence of Cement Composition on the Corrosion of Reinforcement and Sulfate Resistance of Concrete. *ACI Materials Journal*, 87(2):114-122.

Rasheeduzzafar, F.H. Dakhil, M.A. Bader and M.M. Khan. 1992. Performance of Corrosion Resisting Steels in Chloride-Bearing Concrete. *ACI Materials Journal*, 89(5):439-448.

Rasheeduzzafar, S.E. Hussain and S.S. Al-Saadoun. 1992. Effect of Tricalcium Aluminate Content of Cement on Chloride Binding and Corrosion of Reinforcing Steel in Concrete. *ACI Materials Journal*, 89(1):3-12.

Reyna, S.M., J.A. Vanegas and A.H. Khan. 1994. Construction Technologies for Sewer Rehabilitation. *Journal of Construction Engineering and Management*, 120(3):467-487.

Rollings, R.S., J.P. Burkes and M.P. Rollings. 1999. Sulfate Attack on Cement-Stabilized Sand. *Journal of Geotechnical and Geoenvironmental Engineering*, 125(5):364-372.

Sanjuan, M.A. 1997. Formation of Chloroaluminates in Calcium Aluminate Cements Cured at High Temperatures and Exposed to Chloride Solutions. *Journal of Materials Science*, 32:6207-6213.

Sanjuan, M.A. 1999. Effect of Curing Temperature on Corrosion of Steel Bars Embedded in Calcium Aluminate Mortars Exposed to Chloride Solutions. *Corrosion Science*, 41(2):335-350.

Saricimen, H., O.A. Ashiru, N.R. Jarrah, A. Quddus and M. Shameem. 1998. Effect of Inhibitors and Coatings on Rebar Corrosion. *Materials Performance*, 37(5):32-38.

Scanlon, J.M. and M.R. Sherman. 1996. Fly Ash Concrete: An Evaluation of Chloride Penetration Testing Methods. *Concrete International*, 18(6):57-62.

Schupack, M. 1989. Fast Field Test for Chloride Ion. *Concrete Construction*, 34(7):644-650.

Shi, C., J.A. Stegemann and R.J. Caldwell. 1998. Effect of Supplementary Cementing Materials on the Specific Conductivity of Pore Solution and Its Implications on the Rapid Chloride Permeability Test (AASHTO T227 and ASTM C1202) Results. *ACI Materials Journal*, 95(4):389-394.

Sivasundaram, V., G.G. Carette and V.M. Malhotra. 1991. Mechanical Properties, Creep, and Resistance to Diffusion of Chloride Ions of Concretes Incorporating High Volumes of ASTM Class F Fly Ashes From Seven Different Sources. *ACI Materials Journal*, 88(4):407-416.

Sleeper, W. Memo to T. Cavanagh, Metropolitan Water District of Southern California, March 27 1995.

Smith, A. 1989. Concrete Sludge-Holding Tanks Repaired with Shotcrete and Two Colors of Epoxy Coating. *Concrete Construction*, 34(11):947.

Somuah, S.K., J.K. Boah, P. Leblanc, A.-H.J. Al-Tayyib and A.I. Al-Mana. 1991. Effect of Sulfate and Carbonate Ions on Reinforcing Steel Corrosion as Evaluated Using AC Impedance Spectroscopy. *ACI Materials Journal*, 88(1):49-55.

Sontheimer, H., W. Kolle and V.L. Snoeyink. 1981. The Siderite Model of the Formation of Corrosion-Resistant Scales. *Journal of the American Water Works Association*, 73(11):572-579.

Soukatchoff, P, Baron, J., and L. Lessirard. 2001. Cement-based Materials Used in Drinking Water Pipe Systems. *Water Science and Technology: Water Supply*. 1(4):227-236.

Steele, J. 1994. Coating Film Thickness in Concrete Sanitary Sewers: How Much is Enough? *Materials Performance*, 33(9):39-41.

Sudrabin, L.P., P. Kaiser and L. Liszczynsky. 1983. Coatings and Cathodic Protection of Potable Water Storage Tanks. In *Proc. of the NACE International Corrosion Forum*. Anaheim, CA: National Association of Corrosion Engineers.

Suryavanshi, A.K., R.N. Swamy and S. McHugh. 1998. Chloride Penetration Into Reinforced Concrete Slabs. *Canadian Journal of Civil Engineering*, 25(1):87-95.

Swamy, R.N., A.K. Suryavanshi and S. Tanikawa. 1998. Protective Ability of an Acrylic-Based Surface Coating System Against Chloride and Carbonation Penetration Into Concrete. *ACI Materials Journal*, 95(2):101-112.

Szymanski, W.A. and S.J. Scott. 1983. The Use and Performance of RTP in Waste Water Treatment Applications. In *Proc. of the NACE International Corrosion Forum*. Anaheim, CA: National Association of Corrosion Engineers.

Thompson, N.G., K.M. Lawson and J.A. Beavers. 1987. Monitoring Cathodically Protected Concrete Structures with Electrochemical Impedance Techniques. In *Proc. CORROSION/87*. San Francisco, CA: National Association of Corrosion Engineers.

Tikalsky, P.J. and R.L. Carrasquillo. 1992. Influence of Fly Ash on the Sulfate Resistance of Concrete. *ACI Materials Journal*, 89(1):69-75.

Tikalsky, P.J. and R.L. Carrasquillo. 1993. Fly Ash Evaluation and Selection for Use in Sulfate-Resistant Concrete. *ACI Materials Journal*, 90(6):545-551.

Trejo, D., P.J.M. Monteiro, B.C. Gerwick Jr. and G. Thomas. 2000. Microstructural Design of Concrete Reinforcing Bars for Improved Corrosion Performance. *ACI Materials Journal*, 97(1):78-83.

Tseng, T. and M. Edwards. 1999. Predicting Full-Scale TOC Removal. *Jour. AWWA*, 91(4):159-170.

Tseng, T., B.D. Segal and M. Edwards. 2000. Increasing Alkalinity to Reduce Turbidity. *Jour. AWWA*, 92(6):44-54.

Velivasakis, E.E., S.K. Henriksen and D.W. Whitmore. 1997. Halting Corrosion By Chloride Extraction and Realkalization. *Concrete International*, 19(12):39-45.

Velivasakis, E.E., S.K. Henriksen and D.W. Whitmore. 1998. Chloride Extraction and Realkalization of Reinforced Concrete Stop Steel Corrosion. *Journal of Performance of Constructed Facilities*, 12(2):77-84.

von Fay, K.F. and J.S. Pierce. 1989. Sulfate Resistance of Concretes with Various Fly Ashes. *ASTM Standardization News*, 17(12):32-37.

Wee, T.H., A.K. Suryavanshi and S.S. Tin. 2000a. Evaluation of Rapid Chloride Permeability Test (RCPT) Results for Concrete Containing Mineral Admixtures. *ACI Materials Journal*, 97(2):221-232.

Wee, T.H., A.K. Suryavanshi, S.F. Wong and A.K.M.A. Rahman. 2000b. Sulfate Resistance of Concrete Containing Mineral Admixtures. *ACI Materials Journal*, 97(5):536-549.

Weldon, D.G. 2000. Blistering of Epoxy Lining in Potable Water Tank. *Materials Performance*, 39(12):44-49.

Wernham, M. 1995. Epoxy Coating System for Potable Water Tanks and Pipelines. *Journal of the Indian Water Works Association*, 27(3):160-161.

Weyers, R. 1993. *Concrete Bridge Protection, Repair, and Rehabilitation Relative to Reinforcement Corrosion: A Methods Application Manual*. Washington, D.C.: Stragetic Highway Research Program, National Research Council.

Whiting, D. and M.A. Nagi. 1996. Tests for Evaluation of the Effectiveness of Penetrating Sealers in Reducing Penetration of Chlorides into Concrete. In *Techniques to Assess the Corrosion Activity of Steel Reinforced Concrete Structures, ASTM STP 1276.* West Conshohocken, PA: American Society for Testing and Materials.

Yeomans, S.R. 1994. Performance of Black, Galvanized, and Epoxy-Coated Reinforcing Steels in Chloride-Contaminated Concrete. *Corrosion*, 50(1):72-81.

Zhang, M.-H., A. Bilodeau, V.M. Malhotra, K.-S. Kim and J.C. Kim. 1999. Concrete Incorporating Supplementary Cementing Materials: Effect on Compressive Strength and Resistance to Chloride-Ion Penetration. *ACI Materials Journal*, 96(2):181-189.

ABBREVIATIONS

\$/lb.	dollars per pound
\$/MG	dollars per million gallon
µg/L	micrograms per liter
µm	micrometer
µmhos/cm	micromhos per centimeter
µS/cm	microsiemens per centimeter
AL	action level
Alk	alkalinity
ASTM	American Society for Testing and Materials
AWWA	American Water Works Association
AwwaRF	Awwa Research Foundation
CCPP	Calcium Carbonate Precipitation Potential
cm^2	square centimeter
Cond.	conductivity
D.O.	Dissolved Oxygen
DBP	Disinfection By-Products
DIC	Dissolved Inorganic Carbonate
EPA	Environmental Protection Agency
L.I.	Langelier Index
L.R.	Larson's Ratio
L/min.	liter per minute
LCR	Lead and Copper Rule
Max.	maximum
Meq	milliequivalent
mg/L	milligram per liter
MGD	Million Gallons per Day
MIC	Microbially Induced Corrosion
moles/L	moles per liter
IRB	Iron Reducing Bacteria
SRB	Sulfate Reducing Bacteria
mpy	Mils (milli-inches) per year
NOM	Natural Organic Matter
NTU	Nephelometric Turbidity Units
^oC	Degree Celsius
^oF	Degree Fahrenheit
PO_4	Phosphate
ppb	parts per billion
ppm	parts per million
psi	pound per square inch
TDS	Total Dissolved Solids
Temp	Temperature
TOC	Total Organic Carbon

Case Studies
of Treatment Plant Infrastructure Degradation
From Enhanced Coagulation

Marc Edwards
Virginia Tech

CONTENTS

Tables

Figures

Case Study 1: Gothenburg, Sweden

Hosts and Co-authors: Linda Werner, Olle Ljunggren and Olaf Bergstedt

Key Experience:

Rebar corrosion, concrete degradation, epoxy

Background:

The City of Gotheburg has extensive experiences in relation to infrastructure degradation of coagulation, sedimentation and water storage facilities. Both facilities use a combination of pre-lime, alum coagulation, sedimentation and filtration processes. Recently, however, one of the 7 parallel treatment trains at Lackareback has been modified with an innovative dissolved air flotation process in the second (upper) stage of an existing two story sedimentation basin design. The modified treatment process has tended to give about 30-50% improvements in residual aluminum and turbidity, although particle counts appear to be unimproved or slightly higher. Disinfection has always been from a combination of pre and post chlorination, although the Alelyckan plant is currently testing pre-ozonation for part of their flow.

Table 1.1. Summary of Typical Information Relevant to Infrastructure Degradation. Water quality information based on 1999 annual report.

Parameter	**Lackareback**	**Alelyckan**
Constructed	1968	1949
Water Source	River water after detention in Lake	River
Type of Concrete	Original: K 300 Replacement: K 40	350 A
%Water in Concrete when formed	Original: 0.6 Replacement: 0.55	0.55
Raw Water Color (mg Pt/l)	35	20
Raw water Ca^{+2} (mg/L)	6.2	7.4
Average Turbidity (FNU)*	7	0.74
Raw Water pH	7.0	7.2
Raw Water Alkalinity (mg/L as $CaCO_3$)	11.5	13.5
Alum Dose (mg/L)	41.7	37.5
Pre-lime Dose (mg/L)	5	5
Coagulation pH	6.5	6.6
Aluminum Residual (mg/L)	<0.02	<0.02
Final Chloride (mg/L)	14	9
Final Sulfate (mg/L)	27	28

Coagulation and Flocculation:

The most severe infrastructure degradation was noted in the coagulation basin at the newer (Lackareback) treatment plant. In the late 1970's, significant softening of the concrete was obvious, as it could be scraped from the wall with a fingernail, and some signs of rebar corrosion became apparent (Figure 1.1). Conditions deteriorated further until 1995, when a restoration was attempted of the basin walls in one of the seven treatment trains. The general approach was to apply about 20 millimeters of a spray on high density concrete (Densite) throughout the coagulation and sedimentation basis, and 3-4 millimeters of Stocreate were applied in a few hard to reach places. The results were deemed satisfactory and in 1998, restoration of the remaining 7 treatment trains was begun using the same approach. However, at the start of that project, between 10-100 obvious failures of the wall were noted throughout the plant (Figure 1.1). At some points more than 80% of the rebar cross sectional area was eaten away, causing clear breaks in some walls and even allowing water to mix between the parallel treatment trains in some circumstances. Degradation was worse in the first two flocculation chambers compared to the sedimentation basins, perhaps due to the higher water mixing rates or perhaps because the walls were thinner.

Because the underlying concrete clearly did not have the structural integrity necessary for a simple surface restoration, a change in plans was necessary. After consideration of the available options, it was decided to sandwich the old wall between two new walls (Figure 1.2). The anticipated cost of the surface restoration was $150,000 US for each coagulation and sedimentation train, but the decision to install the new walls raised the total cost to $250,000 US. Thus, the overall restoration costs from 1995-1998 totaled approximately $2.5 million US dollars for this 34 million gallon per day (MGD) plant with average flow of 22.4 MGD. The double wall approach had the additional advantage of providing more structural support, which is obviously advantageous if a parallel treatment train was to be drained while leaving another full.

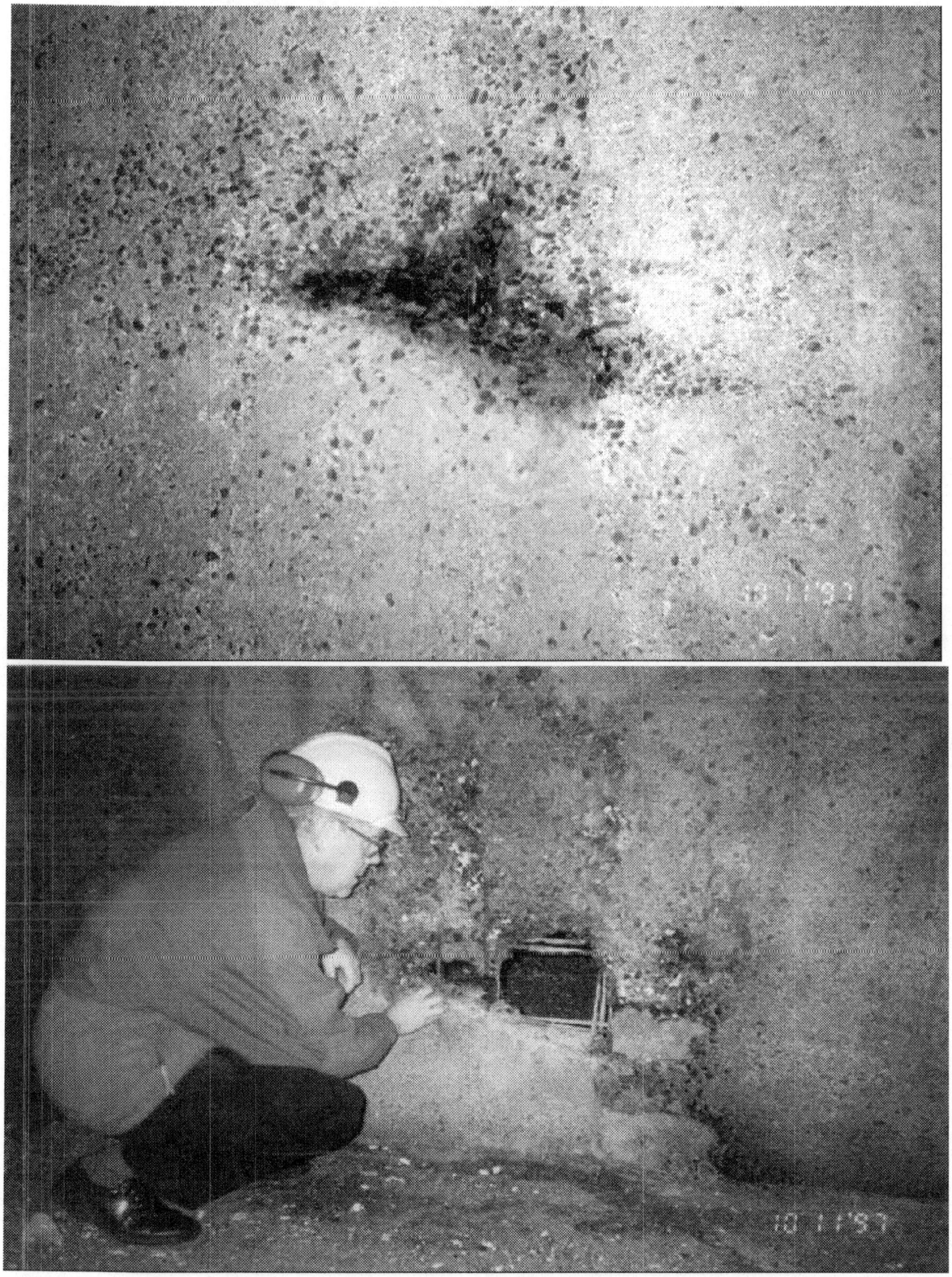

Figure 1.1. Signs of underlying rebar degradation becoming visible on the surface as spots of iron corrosion (above). In numerous instances, complete loss of the wall integrity eventually occurred in the flocculation basins (below).

Figure 1.2. A degraded wall separating adjacent treatment trains was sandwiched between two new walls during restoration.

The most interesting aspect of this case study is that the older plant, which draws water from essentially the same river water source and uses identical treatment processes, has no obvious concrete degradation as of 2000. There are two possible explanations for this, including different construction practices in relation to the concrete and slight differences in water quality. Of these, it is believed that the most important factor was a lower quality concrete at the new plant. The term "lower quality" concrete obviously includes the installation procedures, curing, extent of vibration, in addition to the actual specifications of the concrete used. As evidence of this, when the new plant was repaired in 1995-1998, operators noted that much of the aggregate was at the bottom of the walls. In addition, it was believed that some $CaCl_2$ was added to the concrete at the new plant to enhance curing, although subsequent tests revealed only 0.027% chloride in the concrete by weight, which is below the threshold value of 0.030% believed necessary to cause rebar degradation.

It was considered possible, although highly unlikely, that raw water quality was a contributing factor. The newer treatment plant pumps water from a river to a natural lake reservoir, where it is held temporarily prior to treatment. While in the lake, on average, color tends to increase significantly and turbidity tends to decrease, although there are times during the year where the trends for color are reversed. The alum dosage used at each plant is typically related to the raw waters color, typically 30 mg/L at the newer plant and closer to 40 mg/L at the older plant, but varying throughout the year. However, the typical coagulation pH is 6.5 at both facilities.

Filters and concrete reserviors.

At the newer treatment plant, the filtration basins and the underdrain were showing slight signs of deterioration in 1985-1988, and some "Aquata" Epoxy was applied to the entire inner surface of one filter wall and floor, and only applied to upper part of the underdrains at the 4 other filters. Somehow, application of the epoxy led to detachment of some plastic flow fixtures that set into the concrete of the underdrain. In general, while it is not clear to what extent the epoxy is beneficial, since the exposed concrete is still in excellent condition, this application of epoxy was considered a success.

However, the epoxy was not used for the other filters due to problems experienced in similar environments (see below), and some concerns raised about organics leaching from epoxy contact with drinking water.

The Aquata epoxy was less successful in protecting the inside of water reservoirs. The concrete surface might not have been prepared effectively enough before the application, in terms of surface cleaning or dryness, or some other problem was present. In any case, the epoxy detached from the inner surface of the concrete within 2 years (Figure 1.3), and the utility would not use the epoxy again for this purpose. Other people in the region have had similar problems. Most of the problems with above ground reservoirs were due to external corrosion of concrete supports and the underside of a 0.5 million gallon storage tank (where 4 inches of concrete were lost over 40 years). The concrete on one of 4 reservoirs was recently redone for a cost of about $1.3 million US. The work had to be done 82 feet above the ground, and 90% of the cost was attributable to restoration of the outside surface.

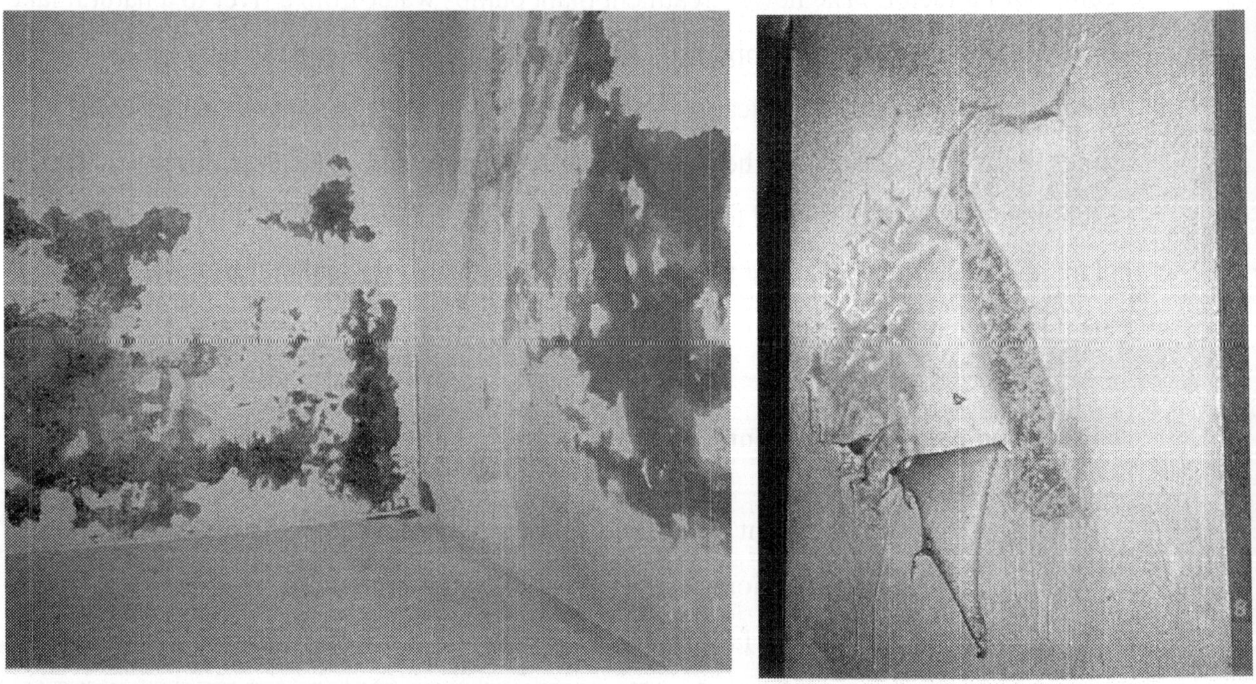

Figure 1.3. Peeling of epoxy from inside water reservoirs was obvious 2 years after application.

Other Observations:

There are no problems with excessive deterioration of pumps and pipes and mixers, although the exterior of some steel pipes was showing signs of visible corrosion, and the plant recently installed dehumidifiers to arrest the problem by preventing condensation. The town has very good experiences with epoxy to prevent deterioration of external steel corrosion.

The site visit highlighted the importance of effective management to stay on top of infrastructure degradation issues and the usefulness of detailed record keeping. Key management at the utility spent about 20% of their time on infrastructure management and oversight of remedial activities. In addition, the town of Gothenburg maintains meticulous records, which is important given that the design life of treatment plants will exceed that of the average operator by about a factor of 10 or so. Old records were recently invaluable during renovation of a reservoir from 1899, when a pipe and other structures were encountered whose original purpose had long since been forgotten. A plant scrapbook is also useful in passing the history of the plant onto new plant personnel.

Case Study 2: City of Regina, Canada

Hosts and Co-authors: Joe Bergman and Ben Boots

Key Experience:

Chlorine Attack on Stainless Steel

Background:

The Buffalo Pound Water Treatment Plant presently has a rated capacity of 37 MGD and supplies water to the cities of Moose Jaw and Regina, Canada. The supply is the man-made reservoir "Buffalo Pound Lake" in the Qu' Appelle River Basin. The treatment plant was built in five stages (Table 2.1). The original plant was brought on-line in September, 1955 and consisted of a pumping station, concrete pipe to the main plant, two 65 ft. diameter clarifiers and four sand filters. In 1958 a pump was added to the pump station and two more clarifiers and four more filters were built. The clarifiers and filters were practically identical to those built in 1952. From 1966 to 1970 improvements were made to the lake pumping station, clarifiers, and filter media. In 1985 eight carbon contactor vessels, two screw pumps, and a reactivation furnace was constructed to improve the treated water taste and odor. The final major expansion was completed in 1989 and included pretreatment facilities, two clarifiers, four filters, and four carbon contactor vessels. The pretreatment facilities consist of cascade de-aeration, flash mixers and four-stage flocculation basins. Two more pumps were installed in the expanded pumping station.

Table 2.1. Summary of phased construction and present age.

Component	Number	Built	Age, yrs
65 ft. diameter clarifier	2	1952	48
Sand Filters	4	1952	48
65 ft. diameter clarifier	2	1958	42
Sand filters	4	1958	42
Carbon contactors	8	1985	15
Screw pumps	2	1985	15
Pretreatment facility	1	1989	11
100 ft. square clarifier	2	1989	11
Filters	4	1989	11
Carbon contactors	4	1989	11

Water entering the main plant from the pump station flows through the pretreatment facility, clarifiers, filters, and carbon contactors and enters the distribution system. The cascades are used sparingly, off more than they are on, and the carbon contactors are used only during the summer and fall months when the blue-green algae growth is high in the lake.

A special focus of this work was four indoor 65 ft. diameter clarifier tank metal works, bridge work that supports the influent cone, a trough, detention shell and collector pipes. When first constructed these were originally steel, some of which still remains on the bridgework of the tanks. However, three of the four steel clarifier tank influent cones, troughs, detention shells and collector pipes were replaced with stainless steel. Thus, the stainless steel clarifier tank component ages were 1, 7, 9, and 11 years at the time of our site visit.

Table 2.2. Age of the clarifier tank steel and stainless steel (SS) components are as follows.

Clarifier Tank	**Built/Replaced**	**Metal/Age, yrs.** **Bridge Work**	**Influent Cone, etc.**
North 1	1952/ 1968	Steel/48	Steel/32
South 1	1952/ 1999	Steel/48	SS/1
North 2	1958/ 1991	Steel/42	SS/9
South 2	1958/ 1993	Steel/42	SS/7
North 3	1989	Galvanized Steel/11	SS/11
South 3	1989	Galvanized Steel/11	SS/11

The plant maintains a written record of maintenance activities and has copies of plant construction plans and specifications. The 1989 construction plans were received and reviewed prior to our site visit and a copy of typical cast-in-place concrete specifications were obtained and reviewed during our site visit. Our site visit consisted of a visual inspection of plant components and a photographic record of observed degradation mechanisms.

With respect to coagulation, the plant has always used alum except for a very short trial with ferric chloride. The water has a pH of 6.8 to 7.4, a sulfate concentration of 150 to 200 ppm, and a chloride content of about 16 ppm. The target effluent free chlorine residual is 1 mg/L, and when GAC is in use, the upstream dose is decreased and the water is rechlorinated after GAC.

Table 2.3. Summary of Typical Information Relevant to Infrastructure Degradation. Water quality information based on personal communication.

Parameter	**Value**
Type of Concrete	Type V
%Water in Concrete when formed	0.45
Raw Water TOC (mg/l)	8-12
Average Turbidity (NTU)	1-20
Raw Water pH	7.7-9.0
Raw Water Alkalinity (mg/L $CaCO_3$)	114-207
Alum Dose (mg/L alum)	55-100
Raw Water Calcium (mg/L)	34-58
Raw Water Magnesium (mg/L)	17-26
Coagulation pH	6.5-7.3
Silica (mg/L as SiO_2)	0.1-7.4
Aluminum Residual (mg/L)	0.0-0.33
Final Chloride (mg/L)	15-16
Final Sulfate (mg/L)	150-200

Coagulation and Flocculation:

Metallic Corrosion.

No degradation of plant infrastructure components was observed that can be attributed directly to the use of the alum. During interviews with plant personnel, no instances were cited by the plant personnel where they were of the opinion that observed degradation was a direct result of the use of alum, other than through normal reduction of pH that accompanies alum dosing.

However, there were two perplexing issues of degradation of infrastructure at the plant, including clarifier fixture anchor bolts that were corroding the stainless steel work on the clarifier itself. The anchor bolts have to be replaced about once every 3 to 4 years (Figure 2.1). Strangely, the 5/8 inch diameter bolts need replacement faster than the ½ in. diameter bolts. The original bolts were 304 stainless steel and more recent replacements have been 316 stainless steel to improve corrosion resistance. The most severe corrosion takes place where the nut screws into the support plate. The chloride content of the treated water in the clarifier tanks is relatively low and the 304 stainless steel should be resistant. However, the chloride concentration within the nut-bolt thread area may be significantly higher then the bulk water. The build up of chloride ions within these areas may initiate corrosion within these crevices, draw further chlorides into the crevices and accelerate the corrosion process. There is also a possibility that the steel was not the specified 304 stainless steel, although a phone call to the manufacturer ensured that Buffalo Pound personnel had every right to believe it was. The call to the manufacturer also confirmed that the ½ inch and the 5/8 inch diameter bolts were of identical composition. Inspection suggested that the only difference between the two bolts was their surface area, diameter, and depth into which they were set into the concrete, with 5/8 inch bolts set into deeper holes. It is anticipated that 316 stainless steel bolts will be more resistant to this type of attack.

The bridgework of the clarifier tanks is painted steel for the North-South 1 and North-South 2 clarifiers, and at about 45 years of age shows little degradation from corrosion. Likewise, the bridgework above the North-South 3 clarifier tanks is stainless steel and at 11 years of age showed little degradation from corrosion. Most of the original clarifier tank influent cones, troughs, detention shells and collection pipes constructed with mild steel have been replaced with stainless steel. Interestingly, both the replacement stainless steel and the original stainless steel in the new clarifier is showing surprising amount of degradation considering its supposed resistant to corrosion and its relatively young ages of 1, 7, 9 and 11 yrs. old (Figure 2.2). The relative corrosion of these stainless steel components appears to correspond with age, with the older elements experiencing significantly more surface corrosion above the water line. In addition, on some of the clarifiers, the bottom horizontal elements (horizontal elements which face the treated water) appear to have more surface corrosion, whereas in the other clarifiers the opposite was true. Also, the corrosion appears to be more extensive around welded areas.

Samples of the rust on the stainless steel where collected and subject to intensive testing that included 1) SEM with ESCA for elemental composition, and 2) dissolution tests with the rust collected from the pipe. The goal was to better understand why the stainless steel degradation was occurring at all. The SEM with ESCA analysis of 4 rust samples determined an approximate elemental composition of 48% Fe, 31% Cl, 12% Cr, 8% Ni and 0.4% Si. This compares to 71% Fe, 0% Cl, 19% Cr, 9% Ni and 1% Si in a typical 304 stainless steel. Thus, the only major difference between the rust and the original metal is the extremely high Cl. When 2 grams of this rust was placed in 20 mls of pure water, the pH was 2.6 after 10 minutes. These results are consistent with attack of free chlorine (Cl_2) gas on the iron in stainless steel near the water line:

$$Fe^{\circ} + 3/2Cl_2 \text{ (gas)} \rightarrow FeCl_3 \text{ (solid)}$$

The $FeCl_3$ that is formed is a strong acid that would further degrade the stainless steel in a humid environment. The steel is corroding to a greater extent above the water line since the acid that accumulates cannot be diluted by the water flow.

It is also suspected that the attack may have been exacerbated by sensitization of the stainless steel during welding of these elements, particularly if they were fabricated with 304 stainless steel. The sensitization of stainless steel occurs when chromium carbide precipitate within non-stabilized austenitic stainless steel. When these steels are heated within the range of 1000 to 1500 °F, solid state reactions take place forming chromium carbide grains. The result is the formation of envelopes of chromium-depleted austenite around carbide, which in certain environments will not be resistant to intergranular attack or intergranular stress corrosion cracking (SCC). Oxygenated environments in the presence of chloride, sulfate, and carbonate increase the tendency for intergranular attack and intergranular SCC.

Concrete Degradation.

A type V Portland cement, which is resistant to severe sulfate environments, was specified in the 1989 construction specifications and is considered typical for construction of plant concrete elements by plant personnel. Sulfate exposures of 150 to 1500 ppm in water is considered only moderate, and Type V cement therefore seems to be over designed from this perspective. However, other reasons for using a Type V cement would be the reduced heat of hydration with the associated reduced thermal stresses. During the first seven days, a Type II cement generates about 80 to 85% of the heat of a Type I cement and a Type V 60 to 75% of a Type I cement. No sulfate attack of concrete elements were observed in the main plant.

Three degradation modes were observed that might be enhanced by the alum coagulant. They are corrosion of concrete surfaces, corrosion of reinforcing steel at shrinkage cracks and construction joint locations, and corrosion of steel components and fasteners. During the site visit the South clarifier tank was drained for inspection. The observed depth of concrete softening is about 4 to 6 mm (Figure 2.3). The rate of penetration is typically a function of the square root of time. For a clear concrete cover depth of 30 mm, it would take well over 100 years to penetrate to the depth of the reinforcing steel. Note, the contractor experienced problems with maintaining the specified clear cover depth of 50 mm during the 1989 plant expansion project. Cover depths were as little as a few millimeters and the reinforcing steel in these shallow cover depth areas corroded within one year after construction. Subsequently, 50 mm of shotcrete was placed over all areas where the cover depth was found to be less than 30 mm. Corrosion of the areas covered with shotcrete in the South 3 clarifier was significantly less than the original cast-in-place concrete which had a specified maximum w/c = 0.45, and this is attributed to the higher density of shotcrete versus the original concrete. As expected, the depth of concrete degradation is greater in areas with higher flow velocities and lower in areas exposed to water only periodically (e.g., the cascade aerators).

Leaking of treated water was observed at concrete shrinkage locations and at construction joints even though water stops were shown on the plans. Observed leak locations were:

- Shrinkage cracking in the West wall of North and South 3 clarifier tanks, on the average the cracks are 6 ft apart, built in 1989. Calcium carbonate has formed in these cracks and stalactites and stalagmites have formed at several locations. At a location behind the high pressure and alum sludge pipes is a relatively large vertical crack and the calcium carbonate is stained an orange-yellowish color. The staining may be the result of the corrosion of the reinforcing steel or pigment of the yellow paint (Figure 2.4).

- Channel wall and bottom along the West wall of the North and South 3 clarifiers, average crack separation distance is 6 ft, built in 1989. Greenish, black precipitant appears at one crack location which may be the result of the corrosion of the reinforcing steel at this crack location (Figure 2.4).

- North end of channel along the West wall of the North and South 3 clarifiers, built in 1989, there is leakage through a crack which appears to be a construction joint between the channel wall and bottom. The crack is about 12 ft. long at this location, is not continuous throughout the remaining section of the joint but the crack appears at several other locations along this construction joint.

- At the North end of the channel where it enters the clarifier room, water is leaking between bottom of the channel and the North end wall. A red precipitant has stained the wall at this location that may be red rust products from the corrosion of the reinforcing steel.

- At the intersection between the channel and the South 3 clarifier West wall, the water leaking through the crack at this location has stained the wall a yellowish-red color which may be from the corrosion of the reinforcing steel or paint pigment. Metals analysis indicated a high concentration of iron in the sample that was collected, consistent with re-bar corrosion.

- Influent channel to Train 1 Clarifier which was built in 1989, there is water leaking through the horizontal construction joint between the channel bottom and wall. Also, there is water leaking through vertical shrinkage cracking in the channel walls.

- Filter influent channel construction joint intersection between the 1952 and 1957 construction stages, water is leaking through this construction joint with staining of the joint areas which may be the result of the corrosion of the reinforcing steel. This channel concrete was placed monolithically, that is, there is not a construction joint between the channel wall and bottom. No water was observed to be leaking through the monolithic concrete at the intersection of the channel wall and bottom. Whereas, when these channels were built with a construction joint between the channel wall and bottom, water leaks through the construction joint even though a water stop was to be placed in this joint.

- Carbon filter inlet channel at the construction joint between the channel bottom and wall. At one location where water is leaking through this joint soundings with a rock hammer indicates that the concrete has delaminated from the reinforcing steel. There is a large amount of red staining of the painted surfaces which is most likely from the corrosion of the reinforcing steel and subsequent cracking and delamination of the cover concrete. Note, construction plans show this joint to be a relative congested reinforcing steel area with a water stop placed in the construction joint. Concrete cover meter testing of this area showed that reinforcing steel was present at the delaminated area.

- Shrinkage Cracking Comments

The visual inspection of the 1952 and 1958 construction of the circular clarifier tanks showed that there was no shrinkage cracking of these older 65 ft. diameter tanks. There was very minor leakage through the interface between form ties and the concrete and the number of form tie zones that were leaking is extremely small.

Whereas the 1989 construction of the square clarifier tanks North and South 3, have shrinkage cracks along the West wall of these tanks. The tanks are about 83 ft. square and thus the West wall is about 166 ft. long. Whereas the South wall, which is about 83 ft. long had very few shrinkage cracks in comparison to the longer West wall. Cover meter testing at several shrinkage crack locations demonstrated that these vertical cracks corresponded with vertical reinforcing steel locations.

Of interest is that the 1989 construction specification specified a type V cement which is to contain a lower amount of C_3A and thus have a lower heat of hydration. The concrete was to have a maximum low case of 0.45, cement content of 576 lbs/cy, a compressive strength 4,350 psi, and a water reducer was to be used in all concrete. However, no concrete shrinkage limits were specified. Although, the w/c is to be considered a moderately low value, the cement content relatively low, and the cement would have a low heat of hydration, the long wall of the 1989 clarifier tank was cracked at separation distances of 4, 6, and 8 ft. In our opinion these vertical cracks which transverse through the wall and extend the height of the wall are shrinkage cracks that are related to a combination of design-construction methods.

Figure 2.1. Photo of stainless steel bolt failures in clarifier.

Figure 2.2. Original painted iron bridgework held up quite well. Stainless steel clarifier is showing a significant degradation. Illustrative light green streaks (within white rectangles placed on pictures) are locations of corrosion products analyzed in this work.

Figure 2.3. Testing of concrete softening depth at bottom of drained clarifier.

Figure 2.4. Illustrative view of rebar corrosion products leaching from concrete and streaking down exterior walls.

Case Study 3: Regional Municipality of Ottawa-Carleton, Canada

Hosts and Co-authors: Ian Douglas and J. Van Den Oever

Key Experiences:
Protective Film from Alum on Concrete, Concrete Degradation by Dissolution of Limestone Aggregate, Sulfate Reducing Bacteria Attack on Iron, Importance of Surface Preparation in Preventing Corrosion of Stainless Steel

Background:

The City of Ottawa maintains two water treatment plants named Britannia and Lemieux Island. The Lemieux Island plant intakes, which were built in 1874, are still in use today. Most of the administrative and maintenance personnel are housed in the Lemieux Island plant. The Britannia plant was built in 1960 and consisted of the intake structure, mixing chambers, settling tanks 1 through 3, and filters 1 through 12. In 1989 settling tanks 4 and 5 and filters 13 through 18 were built. The intake is in the Ottawa River. The river water pH is about 7.0. The water is prechlorinated to 0.5 to 0.7 ppm. After the filters the free chlorine is about 0.1 ppm. Alum has always been used as the coagulant and the water is also treated with about 1 ppm activated silica. It is strongly believed that the activated silica markedly improves particle settling, although our attempts to illustrate this in jar tests were unsuccessful during the day of our visit. The pH of the water is decreased to about 6.0 after alum dosing and then increases to about 6.2 after alum dosing. Lime is added at the end of the treatment plant to raise pH to about 8.2.

No degradation of either the Britannia or Lemieux Island plant infrastructure was observed that could be attributed to use of coagulant at low pH. To the contrary, an unusual film was deposited over surfaces of the treatment plant that were exposed to flocculated water. It is believed that these films had protective properties to the underlying concrete. In addition, degradation of plant infrastructure by concrete, bio-corrosion of iron and steel fixtures, and corrosion of reinforcing steel was gathered.

Protective Film from Alum on Concrete

After draining the slow mix basins for inspection, it was immediately apparent the walls were covered with a film that varied in thickness between 0.05-0.5 cm (0.02-0.2 inches). The film was thickest closer to the point of coagulant injection, and at first glance the film appeared to be an old paint that was beginning to flake off in spots (Figure 3.1). However, the basin had never been painted. In the recent past the surface of the basins had occasionally been sprayed with high pressure hoses, and in areas that had been cleaned most frequently, obvious concrete degradation was apparent. Whether this is because the cleaning process was continually destroying a film with protective properties or whether it was directly due to degradation from the water pressure is not known.

Table 3.1. Summary of Typical Information Relevant to Infrastructure Degradation. Water quality information based on personal communication.

Parameter	Value - Avg. (range)
Type of Concrete	not avail.
%Water in Concrete when formed	not avail.
Raw Water TOC (mg/l)	6.5
Raw Water Turbidity (NTU)	3.7 (1.4 – 40.0)
Raw Water pH	7.3
Raw Water Alkalinity (mg/L $CaCO_3$)	24.5 (17-40)
Alum Dose (mg/L alum)	26 - 34
Raw Water Calcium (mg/L)	8.4
Raw Water Magnesium (mg/L)	2.2
Coagulation pH	6.0
Raw Water Silica (mg/L as SiO_2)	4.0
Aluminum Residual (mg/L)	0.070
Polysilicate Dosage (mg/L as SiO_2)	1.0 – 3.0
Final Chloride (mg/L)	3.0
Final Sulfate (mg/L)	6.5

Extensive attempts were made to characterize this film. After dissolving it in hot acid for analysis by ICP-ES, and making assumptions about the solids present, about 60%

of the total dry mass could be accounted for (Figure 3.2). Aluminum, silica, calcium and magnesium were the major components of this solid. The solid was also carefully examined using x-ray diffraction (XRD). XRD analysis illustrated a series of peaks demonstrating a crystalline nature for the film, and after separating out the peaks that were also present in the underlying concrete and which might have contaminated the sample, a single distinguishing peak remained. This peak did match the dominant peak in the XRD patterns for the solids $Al_2Si_{24}O_{51}$, $Al_6Si_2O_{13}$, and a few other aluminum containing solids. Thus, while a positive identification of the solid phase was not possible using XRD, the data are consistent with deposition of a crystalline aluminum solid phase onto the concrete surface from the water. In an final attempt to identify the remaining unknown constituents of this scale, the sample was examined using ESCA. Spot analysis of the surface suggested a composition of 12-42% Al, 37-55% Si, 19-20% Mg and 3-10% Ca depending on the specific part of the surface analyzed. This result supports the idea that this solid is mostly an aluminum magnesium silicate, and there is some evidence in the literature suggesting that similar solids can form under conditions occurring in drinking water treatment (Kvech, 2001). In a very few spots of the solids, iron concentrations were as high as 91%, but this was anomalous.

The presence of this deposit has significantly impacted the plants decision-making regarding coagulant type. In a 1998 study by Douglas, ferric chloride coagulation at pH 5.5 was proven to have numerous benefits in terms of reduced filter headloss build-up, precursor removal and settled water turbidity compared to alum. Use of ferric was also projected to reduce chemical costs by about 25%. However, a switch to ferric has not been made, in part because it is uncertain whether the lower pH and or switch in coagulant chemical would reduce the likelihood of forming the protective coating on plant infrastructure.

Degradation of Plant Infrastructure by Various Mechanisms

Three relatively minor but still significant examples of infrastructure degradation were noted that might be enhanced by use of coagulant chemicals. These include the corrosion of the concrete and limestone aggregate, biological corrosion of iron and steel fixtures, and corrosion of reinforcing steel at shrinkage cracking locations in the concrete structure of 1989 addition to the Britannia plant.

The Portland cement used in construction of the mixing basins is a very basic substance. Water in equilibrium with the cement has a typical pH of 12-13, thereby making concrete somewhat subject to degradation under acidic conditions. Transport of the concrete is expected to be highest in zones where water flow is high. Mixed in with the cement is an aggregate which is mostly limestone at Britannia and Lemieux Island. This aggregate can also be attacked by acidic conditions. At both locations, the cement was degraded to a much lesser extent than was the aggregate (Figure 3.3). For example, the depth of degradation of the Portland cement is about 2-3 mm and 8-9 mm for the Britannia structures built in 1989 and 1960, respectively. In contrast, the depth of degradation of the limestone aggregate in the 11 and 40 year concrete structures is 3-5 mm and 15-20 mm. In fact, in the older facility the aggregate has completely dissolved, leaving the surface pockmarked in the shape of the aggregate. To some extent the limestone aggregate has acted as a sacrificial material to reduce the rate of attack on the concrete as a whole, but the extent to which use of the limestone is detrimental or beneficial compared to another aggregate could not be established. It was interesting, however, that at Ottawa the aggregate was attacked more quickly than the cement, whereas in most other treatment plants visited the opposite was true due to use of siliceous aggregate that did not dissolve significantly.

Significant corrosion of cast iron and steel pipes and fixtures were noted on occasion (Figure 3.4). The attack was fairly localized and the pipe surfaces were coated with rusted colored nodules interspersed with heavy black deposits. Subsequent testing indicated that sulfate reducing bacteria were present in these nodules at high concentrations, consistent with a rotten egg odor that could be detected when the incrustations were removed from the iron and steel fixtures. Sulfate reducing bacteria are well known promotors of steel corrosion.

Shrinkage cracking has taken place in the 1989 concrete structures (tank walls and channels). The width of the shrinkage cracks ranged from 0.10 mm to 0.50 mm. Water leaking through these cracks contains calcium hydroxide which has been removed from the hardened cement. The calcium hydroxide reacts with the carbon dioxide in the air and forms calcium carbonate deposits on the concrete surface. Leaking of treated water through shrinkage cracks occurs in the lower portions of the tank where head pressures are higher. Higher head pressures are required for smaller crack widths. The shrinkage crack spacing of the North wall of the 1989 settling tanks is about 8 ft. on the average. Corrosion of the reinforcing steel appears to be occurring at some of the crack locations. Again, the shrinkage cracking has occurred and is significantly more severe in the newer concrete tanks (1989 construction)

Other Items.

Minor surface corrosion of stainless steel pipes has occurred at weld locations. It is believed that this corrosion is related to surface preparation methods such as grinding with steel abrasives, coating with pickling material to prevent corrosion of materials that were imbedded in the stainless steel during grinding, or a spray on product also designed to prevent corrosion. At present neither the pickling material or the spray on product is used. Instead the welding personnel, who have been trained at the Sandvik Steel Company welding school, carefully select the grinding materials (brass or stainless steel) and electrodes.

Plant piping is stainless-steel as are other steel fixtures in the settling tanks. The stainless steel is presently 316L whereas 304 was previously used. It is interesting that no corrosion of the stainless components was observed in the Britannia and Lemieux Island plants as had occurred in the Regina plant. This is possibly because the chlorine levels are near the detection limit within the plant at Britannia and Lemieux Island.

Presently, steel ladders in the settling tanks are being replaced with fiber reinforced polymer (FRP). The ladder components are bolted together and to the concrete walls with stainless steel bolts (type is not presently known). Care needs to be used in the selection of FRP materials. If the fiber is carbon rather than glass, a galvanic corrosion cell can form between the carbon and stainless steel. The stainless steel bolts would be the anode and the corrosion reaction would be relatively rapid with a large cathode (carbon fiber) in contact with a smaller anode (stainless steel bolt). Also, if the FRP fiber is glass and the polymer is vinyl ester the strength of this FRP in water is reduced by 10% in a month and by about 30% in a year. Strength reduction after 1 year is presently not known.

The Ottawa water treatment plants began using a computer based maintenance management system in 1984. Since 1989 they have been using the MAP COM software and are presently in the process of selecting another software package. Plant personnel stated that the percentage of their maintenance breakdown work decreased from 40% to 5% with the implementation of their preventative maintenance program. All preventative maintenance work is scheduled through the computerized preventative maintenance program. Their working philosophy is that the job is not complete with the "tightening of the last bolt but with the completion of the work order paperwork."

Figure 3.1. Documentation of an unusual film depositing on pipe and pump surfaces contacting water (above left) and concrete mixing basins (above left). Deposits were heaviest in locations immediately after coagulant dosing, but were significant throughout the plant and actually appeared to be an old layer of paint at first glance (bottom left). In some locations the deposits were thicker than a penny, with a yellow white color on the side facing the water and a plant color where contact was made with the underlying cement (lower right).

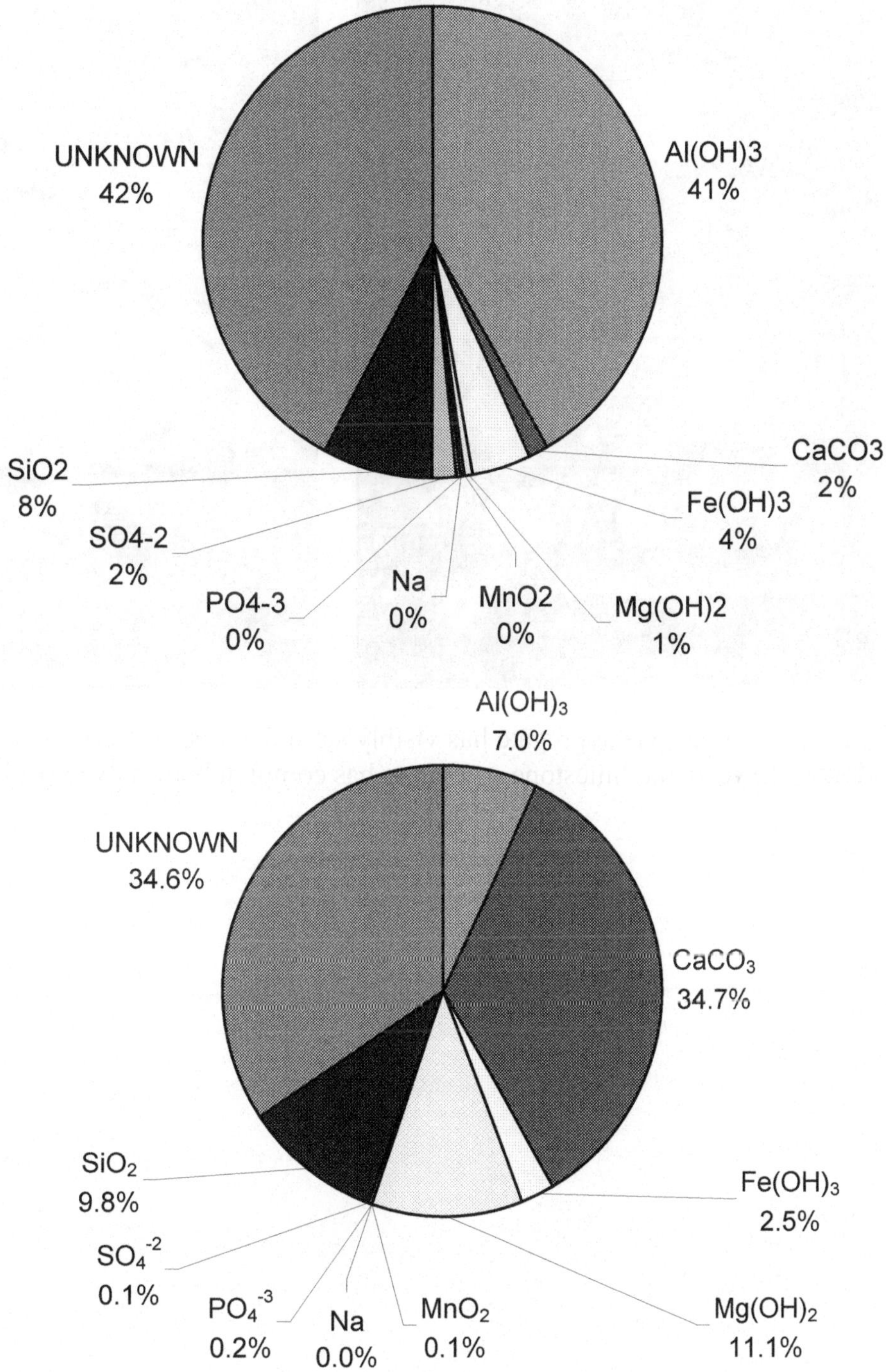

Figure 3.2. Composition of deposits collected from Lemuix Island (above) and Britannia (below) treatment plant surfaces as determined after acid digestion and ICP-ES.

Figure 3.3. Limestone aggregate has visibly started to dissolve after a few years (left), and after 40 years the limestone aggregate has completely dissolved (right).

Figure 3.4. Failure of an infrastructure component after a long period of use. Subsequent analysis determined sulfate reducing bacteria had been active on the metal surface.

Case Study 4 and 5: Billings and Lockwood, Montana

Hosts and Co-authors: Boris Krizek (Billings) and Rick Russell (Lockwood)

Key Experience:
Cl_2 attack on Metals, Copper Pitting Corrosion, Discolored Water, Ferric versus alum

Background:

The Billings and Lockwood treatment plant are located in adjacent towns in Montana. Both plants draw essentially the same water from the Yellowstone River, with Lockwood located about 5 miles downstream from Billings. Site inspections were conducted during the same week in October of 2000. It was anticipated that a comparison of these two plants using identical water would prove instructive. Results of the Billings site visit are discussed first and followed by Lockwood.

Table 4.1. Summary of Typical Water Quality Information Relevant to Case Study.

Parameter	Billings	Lockwood
Date Constructed	1950	1986
Type of Concrete	Unknown	Type I/II
Typical Flow (MGD)	20	1.1
%Water in Concrete when formed	Unknown	49 max
Coagulant History	45 years of alum until 1995, then $FeCl_3$	Alum until 1998, then $FeCl_3$
Water Source	Yellowstone River	
Raw Water Color (mg Pt/l)	NA	
Range Turbidity (FNU)*	3-4000	
Raw Water pH	8.2	
Raw Water Alkalinity	149 mg/L as $CaCO_3$	
Silica (mg/L as SiO_2)	12 mg/L as SiO_2	
Raw Water Calcium (mg/L)	20-44	
Raw Water Magnesium (mg/L)	6-14	
Coagulation Specifics		
Coagulation pH	7.2	6.8-7.1
Finished Water Alkalinity (mg/L as $CaCO_3$)	120	120
Finished Water Cl_2	0.93	1.4
Final Chloride (mg/L)	16 mg/L	
Final Sulfate (mg/L)	24 mg/L	

Site Investigation: Billings Water Treatment Plant, Billings Montana.

A tour of the plant was conducted on the afternoon of one day and the plant inspection a day later. Plant inspection consisted of visual observations and interviews with plant operators. Initial discussions indicated that the Billings plant had problems with failure of copper plumbing. In addition, during the site visit it became clear that fairly severe corrosion of steel in the filter building was occurring. Our site visit attempted to diagnose these two problems.

Copper Pitting Problem

After the alum feed was stopped in May 1995 copper tubing in the plant began to fail at a frequency of about two leaks per month. Prior to that time, copper was used successfully for the chlorine feed line, water feed to polymer tanks, and most other routine plumbing applications. Perhaps most dramatically, a set of copper cooling coils that had been in service for about 19 years failed just 9 months after the alum coagulant was switched to ferric, and the brand new replacement coils also failed 9 months later. Thus, it is clear that some sort of change had occurred that was detrimental to performance of pure copper in the plant.

To solve this problem copper plumbing is gradually being replaced in the plant by plastic, and the copper cooling lines have been replaced with a more expensive nickel alloy that has not yet failed. However, other changes have also been made in the cooling system including less frequent use, implementation of airflow for cooling at certain times of the year, and lower water velocities in the system.

Among the ideas that have been forwarded by plant personnel to explain the pitting problem are:

1) Reduction in effluent pH sufficient to disturb protective $ZnPO_4$ or other coatings. On occasion in 1995, effluent pH dropped as low as 6.5 before a caustic feed was installed to maintain pH above 7.2

2) High pressure in lines and erosion corrosion. Calculated flow rates in the ½" copper were 13-20 ft/sec when the pressure was 70 psi and many failures were at elbows and bends. Reducing the pressure to 35 psi, which was estimated to reduce flow to about

9 ft/sec, did reduce failure rates significantly. On the other hand, much higher pressures were used at the plant prior to 1995 without problem.

3) Gradual reduction in $ZnPO_4$ inhibitor dosing gradually from its start in 1980 until dosing was stopped in about 1989.

A representative sampling of failed pipe materials was collected from a waste pile at the plant (Figure 4.1 and 4.2) and subject to detailed analysis. The general approach was to carefully examine each sample by eye and take normal photographs of any key features. Thereafter, the surfaces were scrutinized by scanning electron microscopy. Finally, ESCA was used in conjunction with SEM to identify elements with atomic weights higher than oxygen. The instrument used is a CamScan Series 2 (CamScan, Cranberry Township, PA) with a quantum system 4000 EDS (American Nuclear Systems, Knoxville, TN).

Typical pits through the copper were about 1 mm in diameter and tubercles were completely absent (Figure 4.3 and 4.4). Under SEM, one of the most important observations is that the copper surface is mostly bare and shiny, with very few deposits on parts of the surface that might offer protection to additional corrosion. Traces of chloride, iron and silica were present (Table 4.2), but in general their concentration was very low. In contrast, no copper could be detected at points of the surface on the copper water meter component, which was heavily encrusted with iron, calcium and silica.

Given these analysis it is clear that a protective $ZnPO_4$ coating, if it was ever present from inhibitor dosing, is now completely absent. No traces of Zn or phosphorus could be detected on any of the pipe surfaces. The bare shiny copper metal on the pitted samples offers strong evidence for erosion corrosion. This is consistent with the reduced failure frequency observed when flow rates were decreased down to 9 ft/sec. Likewise, cupro-nickel tubes are much more resistant to this type of corrosion than is pure copper (Obrecht et al, 1960).

One way to completely avoid erosion corrosion is to make sure that the velocity in pipes is always less than 6 ft/second in cold water lines and 1.5 ft/sec in hot water lines (AWWARF, 1996). If conditions are "favorable," much higher velocities can be endured. What conditions lead to a favorable condition? While a detailed assessment is not available, in isolated incidences it has been noted that lower pH, more dissolved oxygen, particulate matter in water, higher temperature, continuous flow, and larger tube diameter are all unfavorable (AWWARF, 1996; Knutsson et al., 1972). No mention is made of the impact of phosphates, although phosphates generally have a beneficial impact on the rate of copper corrosion.

In summary, the available data suggests that a water quality change occurred in 1995 that made the water more susceptible to erosion corrosion. The lower pH, increased Cl^- versus SO_4^{-2} when changing from alum to ferric coagulants, along with the gradual elimination of phosphate inhibitor are believed to be the most important changes that could have made the water more susceptible to this phenomena. To prevent additional problems, the plant is on the right track in reducing velocities within the copper pipes, or switching to copper nickel or other plumbing materials. If desired, it would also be possible to conduct on-site experiments to determine which specific changes in plant water quality were responsible for increased susceptibility to erosion corrosion that occurred in about 1995. However, the cost of this assessment might exceed the possible benefits to the plant.

Corrosion of Steel in the Filter Gallery

During the on-site inspection three prominent examples of steel degradation were observed including 1) degradation of an older wall in the pump house, 2) corrosion of structural steel members in the filter metal building and the filter annex metal building, and 3) cracking of the concrete flume in the filter metal building.

The filter building was built about 1966 and its annex in 1976. Thus the buildings were about 34 and 24 years old at the time of our visual inspection. The main filter building structural steel (Figure 4.5) members has two coats of paint, a prime and a topcoat. Both coats of paint were most likely applied prior to erecting the structure. The filter building annex is a T shaped metal building with two wings at the end of the main structure. The interior height of the annex is about 10 feet whereas the filter building height is much larger and of 20-30 feet in height. Thus, there is a significant difference in the enclosed volume space between the filter building and its annex. Also, the structural steel members in the annex only have primer and no paint.

Inspection of the steel members indicated superficial corrosion for materials used in the main building, but severe corrosion in the members associated with the annex, or even in steel within the main structure but installed as a vertical member at the entranceway between the filter building and the annex. In fact, a few of these members were no longer fastened to the concrete floor because the steel had corroded completely away (Figure 4.5). Likewise, several sections of steel within the annex wings were nearly completely consumed by corrosion.

The concrete flume in the filter building was discovered to be cracked and leaking. The cracking may be shrinkage and structural cracking. The foundation of the flume may have moved and cracked the flume. Treated water is leaking out of some of the cracks and some of the cracks have been plugged with a calcium silica compound (Table 4.2). At least one very small leak spot appears to be surrounded by corrosion products from reinforcing steel (Figure 4.6).

One obvious cause for the more severe attack on steel in the annex was that only a single coat of primer paint was used. Comparing the two buildings, it is obvious that the additional paint in the main structure almost completely eliminated severe corrosion. However, what was the cause of the attack? Water flows in the flume under the structure, and several vents connect the flume chamber to the structure, as well as the surface of water above the filters. The smell of chlorine was noticeable in the air of the structure, and because it was enclosed humidity was relatively high. As was the case at Buffalo Pound, where it was hypothesized that free chlorine gas caused attack of stainless steel above the water line, the same mechanism is believed to be operating at Billings. The difference is that the normal steel was much more aggressively attacked. In support of this hypothesis, rust from the corroded beam was comprised to 96% Fe and 4% Cl (weight percent of elements with mass greater than oxygen) on ESCA (Table 4.2). Nearly undetectable Cl was found in rust collected from a room with lower smell of Cl_2. When 2 grams of this rust was placed in 20 mls of pure water, the pH was 3.4 and 3.6 after just 10 minutes, consistent with the formation of iron chloride salts and a low pH film on the steel after attack from free chlorine.

Another aspect of this type of attack related to enhanced coagulation is the importance of pH in formation of Cl_2. That is, transport of Cl_2 gas from water becomes much more significant at lower pH, and is insignificant at higher pH. Consequently, steel in buildings housing filters or other waters containing free Cl_2 might be an unforeseen casualty of lower coagulation pHs.

Table 4.2. Analysis of solids and failed copper pipe samples collected from Billings water treatment plant. Elemental composition representative of typical results obtained from more than three measurements of the indicated area on each sample.

Sample Description	Elemental Compostion	Comments
Copper Tubing Failures		
Cooling tube (Fe/Cu joint)	Cu Surface: 99.5% Cu, 0.5% Ca Fe Surface: 100% Fe Rust Deposit: 91% Fe, 9% Si	Very clean copper
Copper Pit on Cooling Tube	Pit: 83% Cu, 11% Cl, 1% Fe, 2% Si, 3.3	Some chloride in pit
Newly Pitted Copper	Solder: 92% Sn, 4.5% Si, 2.5% Fe, 1.4% Cl Pit and nearby surface: 100% Copper	Very clean copper
Pitted Copper Tube	Yellow Streaky Deposits: 97% Cu, 1% Cl, 2% Si	Very clean copper
Pitted Water Meter (Distribution system)	Pitted Area: 56% Fe, 25% Ca, 13% Si, 2% S, 4% In Pits: 92% Cu, 8% Fe	Heavy iron deposits
Failed Beam Samples		
Rust from Corroded Beam	Room with higher Cl_2 and failed beam: 96% Fe, 4% Cl Room with lower Cl_2: 87% Fe, 11% Si, 2% Si	
Cracked Flume Samples		
Thick External Deposit on Cracked Concrete	White Precipitate: 37% Ca, 61% Si, 3% K	

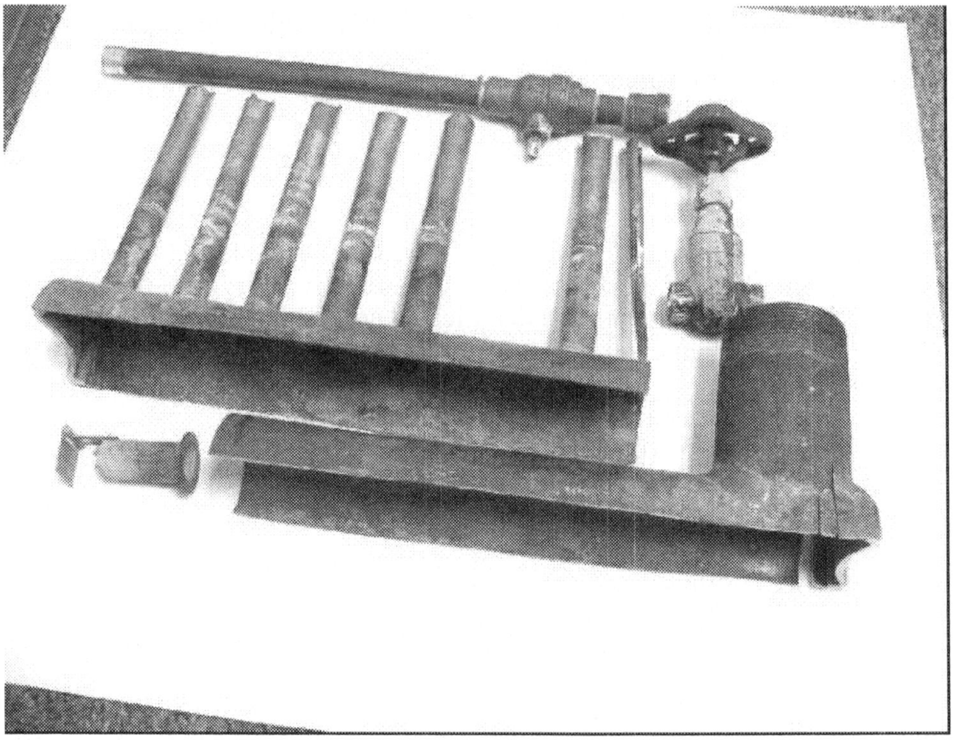

Figure 4.1. A variety of copper tube samples that had failed from pitting were collected for detailed analysis.

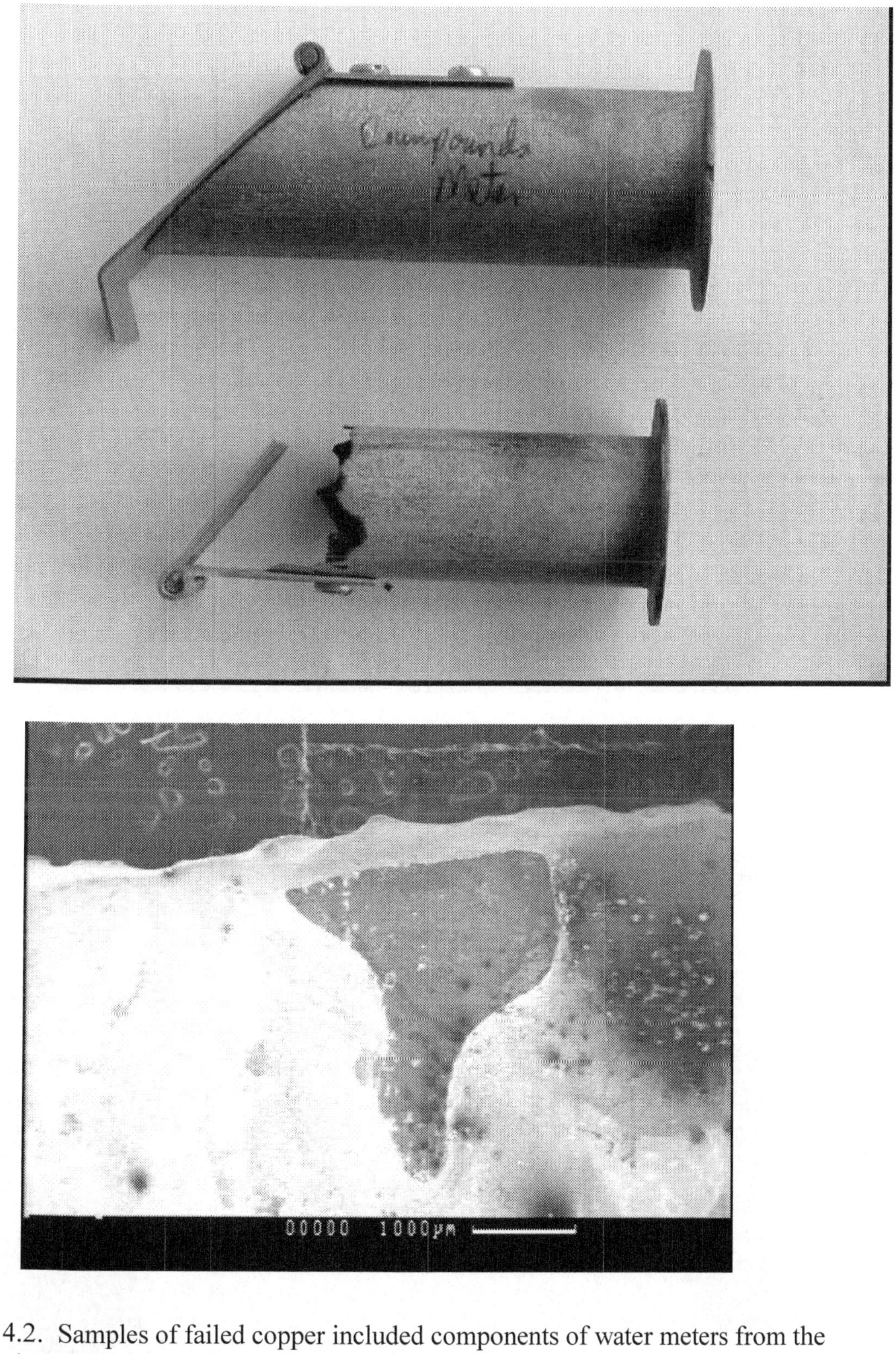

Figure 4.2. Samples of failed copper included components of water meters from the distribution system (above). All samples were subject to scanning electron microscopy with ESCA to determine surface elemental composition.

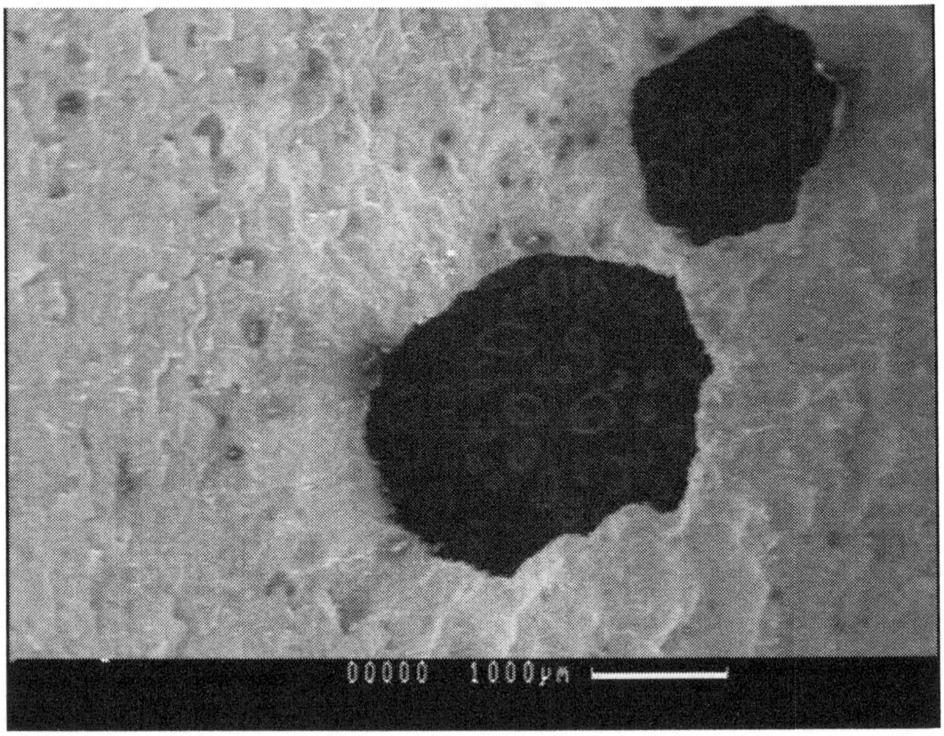

Figure 4.3. SEM photograph of pit on a cooling tube.

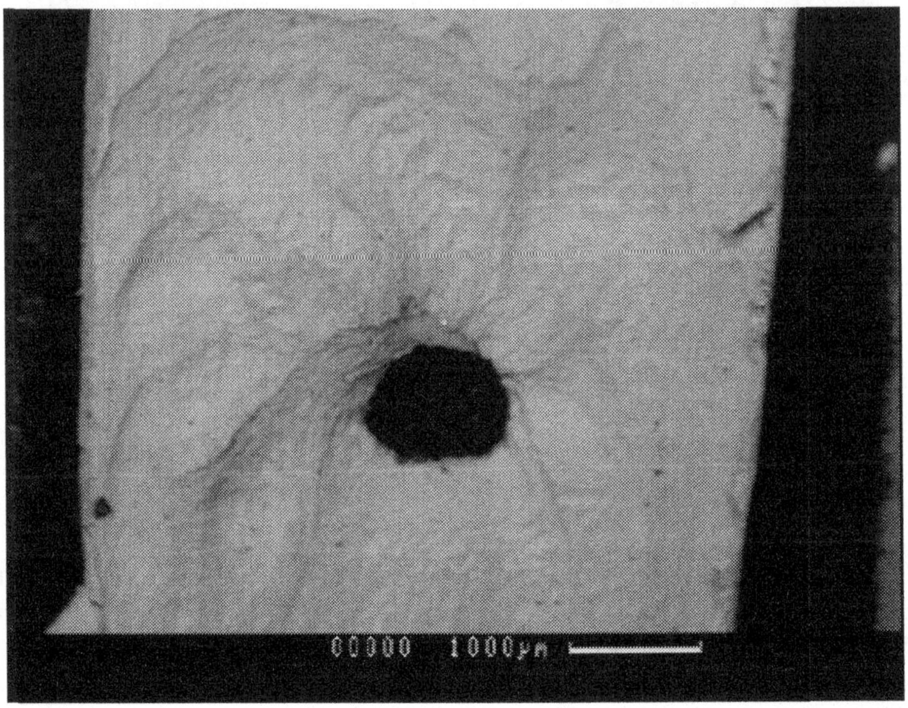

Figure 4.4. SEM of a fresh pit on a new copper pipe.

Figure 4.5. Phenomenon of Cl_2 gas attack on steel beams as exhibited in filter gallery in Billings. Rusting of unpainted beams is clearly apparent on ceiling of structure (above), and many structural beams are nearly completely eaten through in the main room (lower left) and in an adjacent room (lower right).

Figure 4.6. External cracking and deposit formation on concrete flume to filters.

Site Investigation: Lockwood Water User Association Water Treatment Plant, Lockwood Montana

The key issues to be addressed at Lockwood included an overall inspection of the plant, non-destructive testing of key structural components, examination of discolored water that was causing staining of fixtures at the plant and in consumers homes, and discussion of increased problems with copper leaching at the new homes in the distribution system.

Overall Inspection of the Facility

Plant personnel retained a copy of the construction plans and these were sent to the inspection team prior to the visit. Design plans specified a concrete mixture of 4000 psi compressive strength at 28 days, air entrained, 564 lbs of Type I/II Portland cement, and maximum water to cement ratio (w/c) of 0.49. By today standards and in 1980, a w/c of 0.49 would be considered a very high standard for this type of construction. Inspection of all components of the plant by personnel is performed on a regular basis and detailed maintenance records are maintained in a paper format.

From startup to about 1998 years alum was used as the coagulant, but $FeCl_3$ was used thereafter. There were some problems with alum precipitating out of solution and clogging pipes, but with the plant change over to $FeCl_3$, the plant has not experienced similar problems. The Type I/II cement has not seen any noteworthy degradation from the water that is treated or the coagulant chemicals in use.

The main problem noted at the plant is a relatively large amount of shrinkage cracking (Figure 5.1). In our opinion, the relative large amount of shrinkage cracking is the result of the relatively high specified w/c of 0.49. Treated water leaking through cracks leach out calcium hydroxide which converts to calcium carbonate when exposed to air and forms a significant deposit. It is possible that the additional chloride in the water from the $FeCl_3$ is contributing to this problem, but given the levels of Cl^- in the water we think this is only a small contributor. At some locations, some corrosion of rebar was apparent, and the resulting leaks should be closely tracked, especially that observed in the concrete clarifier tank interior wall (Figure 5.1). The leak started as the result of shrinkage cracking and/or lack of consolidation of the concrete at the time of construction, and has been noted to be present over the last five years. Green $FeCl_2$, black $Fe(OH)_2$, and red $Fe(OH)_3$ corrosion products were observed at the leak at increasing distance from the leak in the order presented, and the presence of iron was confirmed by examining a sample with ESCA.

Shrinkage cracking was also quite pronounced in the exterior walls (exposed building wall) of tanks holding filter backwash water, the clarifier and the clearwell (Figure 5.1). Treated water is leaking through some of the cracks and some of the cracks and a few have been plugged by the deposition of calcium carbonate. No iron corrosion products were observed at any of these crack locations and so attack on the rebar is not apparent. The distance between the shrinkage cracks is about 4 feet on the average and appears to extend the height of the reclaim and clarifier tank walls. Efforts to patch the leaks in the clearwell by patching have failed because of poor application procedures and poor choice of materials.

There are several valve corrosion problems that may have been accelerated by the switch from alum to ferric chloride. The backwash pipe values have a cast iron wafer (Figure 5.2), and corrosion of small holes at wafer-pipe contact point caused valve failure. The corrosion is typical crevice corrosion which is accelerated in the presence of chloride ions. The problem is gradually being resolved by replacing the cast iron wafers with bronze wafers. The other valve corrosion problem is in a pressure equalizer piston valve. Within the valve, a horizontal piston seals by contact against a leather gasket. Corrosion products have collected at the bottom of the piston and worn away at the gasket, thereby preventing the valve from closing properly. It is currently believed that if the valve piston was mounted vertically, failure would not occur. In addition, a significant fraction of the plants Dezurik valves have failed, but this is believed to be attributed to a failure to manufacture to specifications rather than any aspect of coagulant selection.

Another observed degradation problem that might be attributed to the ferric was a general attack of a concrete floor section by treated water (Figure 5.3). The aggregate was clearly apparent, and this was attributed to an acid base reaction between the treated water (pH of 7.6) and the concrete with a pH of about 12-13.

Other experiences noted at the plant include a very wide (1/8 in) crack was observed in the slab-on-grade in the hallway outside of the ferric storage room. The crack is perpendicular to the hall wall and extends the width of the hall before continuing underneath a block wall of the ferric storage room. There appears to be an elevation change across the crack and the crack width and elevation change across the crack appears to have remained the same for a long time period. The crack is most likely related to soil settlement.

Non-Destructive Testing

Non-destructive testing was conducted at the following locations:

- At hallway floor slab crack location adjacent to the ferric storage room, ultrasonic wave velocity to assess the quality of the slab on grade concrete and crack depth and rebound hammer to estimate the compressive strength of the concrete.

- Rebound hammer at four wall locations (North, East, South, and West side) at about 4 ft. above ground level on the exterior of the clearwell tank wall. Soundings at select shrinkage crack locations.

- Rebound hammer at one location on the South wall of the reclaim and clarifier tank exterior walls at about 4 ft. above the ground surface. Soundings at select shrinkage crack locations.

The following presents a summary of the above non-destructive test results. The crack in the floor slab is about 0.07 inches wide. The concrete is of good quality with an estimated compressive strength of about 5,500 psi. The crack depth is about 6 inches which most likely is the thickness of the slab. The clearwell tank wall concrete is relatively uniform with average rebound numbers of 47, 39, 40 and 40, North, East, West, and South locations, respectively. The average estimated compressive strength is about 5,500 psi. No delaminations were detected at the sounding shrinkage crack locations when sounded with a 3 lb. hammer. The south wall concrete of the reclaim water and clarifier tank is relatively uniform with rebound numbers of 48 and 47, respectively. The estimated average compressive strength is about 5,500 psi. Hammer soundings at shrinkage crack locations indicated that there are no delaminations at the test locations.

In summary, the observed visual degradation bears frequent monitoring, but has not yet impacted the structural integrity of each location.

Precipitate in water heaters and on plant monitoring equipment.

After the plant switched from alum to ferric, view ports to rotometers and other instruments have become stained and several consumers have also complained about deposits in hot water heaters. There are two likely causes for this. The first is Mn^{+2} contaminants in the $FeCl_3$, which can carryover through the plant and precipitate as a dark compound in the water distribution system. These types of problems can best be addressed by controlling the concentration of Mn^{+2} in the coagulant. The second likely explanation is residual ferric solids in the treated water. A sample of the brown precipitate that coated the plant instrumentation was carefully collected, digested in strong nitric acid and analyzed for metals by ICP-ES. This analysis indicated a metal content of 99% Fe, only 1% Mn and 0.1% As by weight. Thus, the staining is apparently caused by iron and not manganese. The staining can be minimized by examining strategies to reduce effluent iron carry over through the filters, but in the absence of this, removal of stains can be achieved more easily using commercially available cleaners designed to dissolve Fe deposits.

More Persistent and Higher Levels of Copper Leaching in Homes

Perhaps the biggest concern related to changing coagulant at the plant is that new homes have started to see significant problems with relatively high (≈ 6 mg/L) copper levels at the consumers tap. These levels of copper release have been sustained for months and months. In the past, copper levels at the tap in new homes very quickly dropped off to levels that were well below 1 mg/L. As was the case with the copper pitting problem observed at the Billings plant, there are a wide variety of potential causes for this problem. After careful discussions with plant personnel, in this particular case the phenomena we believe most likely to be responsible is known as "hindered aging" of the copper pipe. To understand this problem, it is first necessary to review results of some recent laboratory experiments conducted at Virginia Tech.

When water is allowed to sit stagnant within an aerobic copper pipe, the scale layer on the pipe wall will slowly dissolve until the water is holding all the copper it can. The holding capacity is the maximum level of copper that can be dissolved in the consumer's drinking water at a given time. The identity of the scale (oxidized copper rust layer) that is on the pipe wall and in contact with the solution is an important factor controlling the maximum level of soluble copper. At least 20 different compounds of copper might form on a pipe wall under various circumstances-- each has its own color and maximum level of soluble copper. At the extremes, certain black copper sulfide scales (CuS) on pipes are so insoluble that the maximum level of copper in solution is below the detection limit, whereas fresh light blue copper hydroxide solids [$Cu(OH)_2$] are so soluble that tens of mg/L soluble copper can be dissolved into the water. The goal of water treatment for copper corrosion control is to replacing more soluble solids on the pipe wall with less soluble solids.

Thankfully, it is a general rule of chemistry that solids are destined to change from more soluble to less soluble forms with time-- we term this phenomenon "solids aging." Solids aging can be readily observed by placing fresh cupric hydroxide solids into a beaker and measuring soluble copper with time (Figure 5.3). Over a period of days in the laboratory, the blue $Cu(OH)_2$ solid will gradually change to tenorite with loss of water:

$Cu(OH)_2 \rightarrow$	$CuO + H_2O$
"cupric hydroxide"	"tenorite"
relatively soluble, light blue	relatively insoluble, brown

Because transitions of this nature also occur with time in pipes, levels of soluble copper in old pipes are often much lower than in newly installed pipes. To illustrate, in a water at pH 6.6, alkalinity of 60 mg/L as $CaCO_3$ and 10° C, if a $Cu(OH)_2$ scale is on the pipe wall and controlling release the predicted maximum soluble copper in drinking water will be about 30 mg/L (Figure 5.4). In contrast, for an older pipe in which the $Cu(OH)_2$ scale has been replaced by CuO scale, the maximum soluble copper is only 0.8 mg/L. Thus, maximum levels of soluble copper observed in a home can be expected to decrease by factors of 10 or more with time. In addition, as Figure 5.4 indicates, if a change from $Cu(OH)_2$ to CuO or a less soluble solid does not occur, the pipe will continue to leach very high levels of copper.

The rate of aging depends on the exposure conditions and the water quality. In the example cited in Figure 5.4, the transition from $Cu(OH)_2$ to tenorite was essentially complete in the laboratory after just a month at pH 7.0 at 23° C. However, because other constituents in the water such as silica and natural organic water can interfere with aging, in real situations the aging process might be complete in a few seconds or it might not occur at all. The most important factor the treatment plant could do to increase the rate of aging is to increase the concentration of OH^- ions in water (as measured by the pH). For example, recall that if fresh $Cu(OH)_2$ solids are placed in a beaker at pH 7.0, in a few days they are begin conversion to the less soluble CuO solids by aging and the transition is complete after a few months (Figure 5.5). However, at pH 7.5 and pH 9.5, this transition is complete after a few weeks or hours, respectively. At pHs well below 7.0 the transition sometimes does not occur even after a few years.

In relation to the specific problem at Lockwood, when the change was made from alum to ferric the typical pH of the treated water decreased from about pH 7.0 to pH 7.3. The minimum treated water pH is now as low as pH 6.3 versus about 6.7 before the change in coagulant type. We believe that this reduction of about 0.3 units in pH, which translates to a 50% reduction in the concentration of OH^- in the water, is responsible for the persistent copper release observed in the Lockwood system. If so, installing a caustic or lime feed at the plant to raise pH back to the levels more typically observed at the plant should solve the problem.

Summary:

One of the more interesting results of this study is that the key infrastructure problems were so dissimilar between the two cities. The major factor for this is probably the differing age of the plants and materials in use. However, it was also interesting to note that the corrosivity of the water after it left the plants were so different, based on key corrosion problems that occur in the distribution system. In one case a pitting problem was occurring, yet in the other city the problem was with higher levels of copper in new homes. Among many other possible factors, coagulation conditions and inhibitor usage might contribute to the type of problem that is observed (or avoided).

Figure 5.1. Shrinkage cracks in clearwell allowing slow leaks to occur.

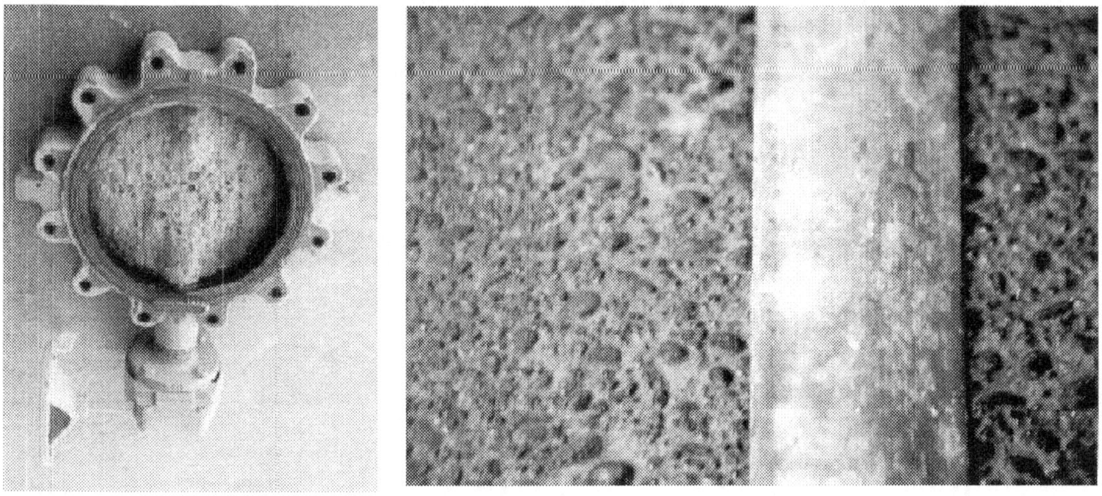

Figure 5.2. Cast iron attack in valves and some etching of a concrete floor are about the only significant problems of infrastructure degradation at Lockwood that might be attributed to a change from alum to ferric.

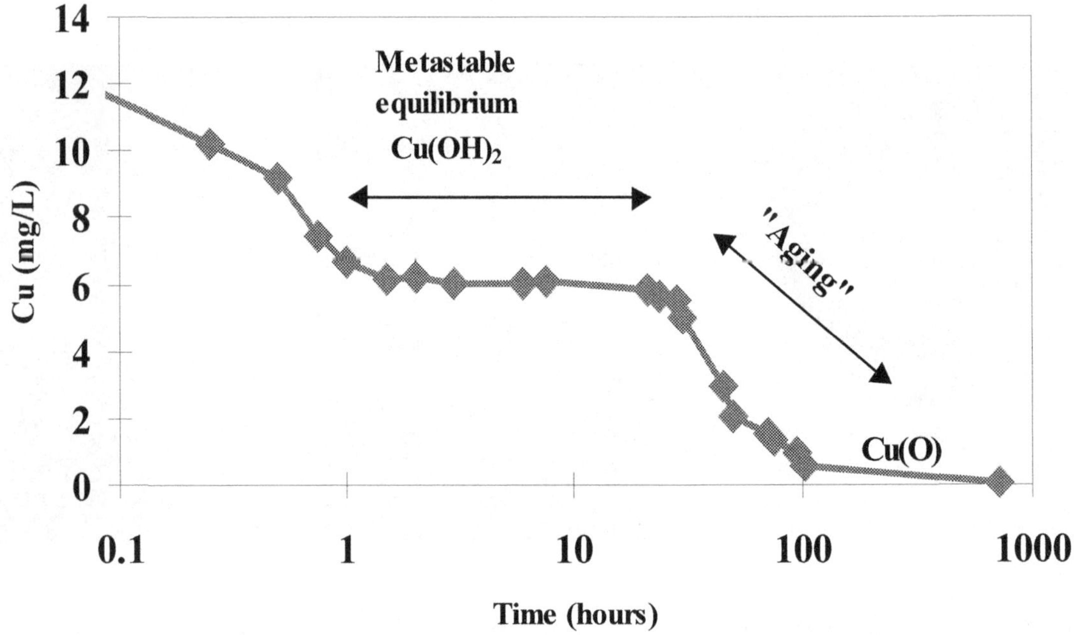

Figure 5.3. As solids age, they can become much less soluble. The rate of aging depends on many factors.

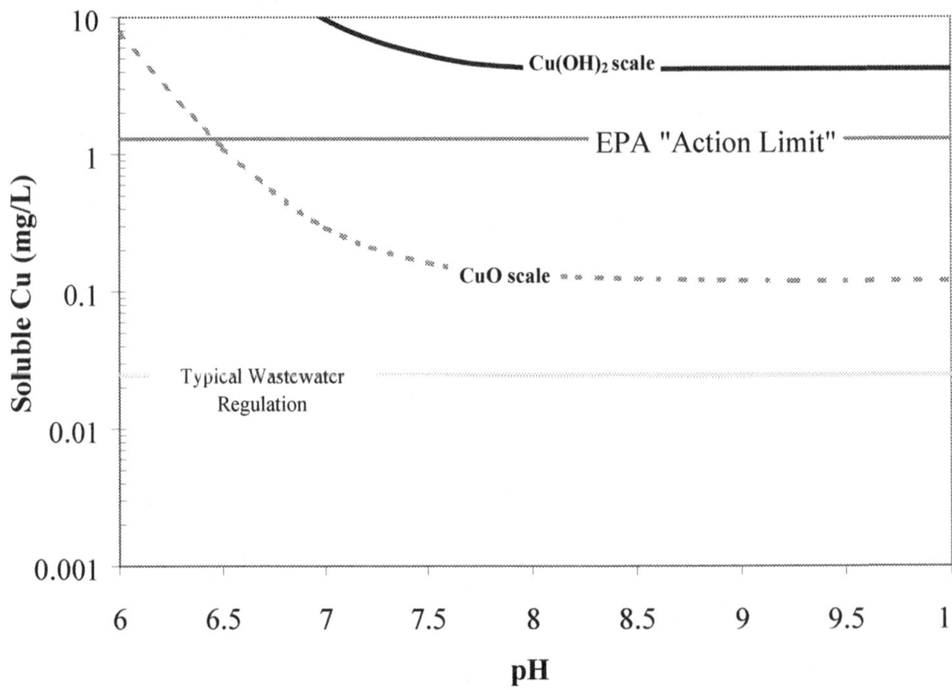

Figure 5.4. Predicted soluble copper as a function of pH in the presence of two solids typically found on pipe walls. Model conditions: 10° C, alkalinity = 60 mg/L as $CaCO_3$.

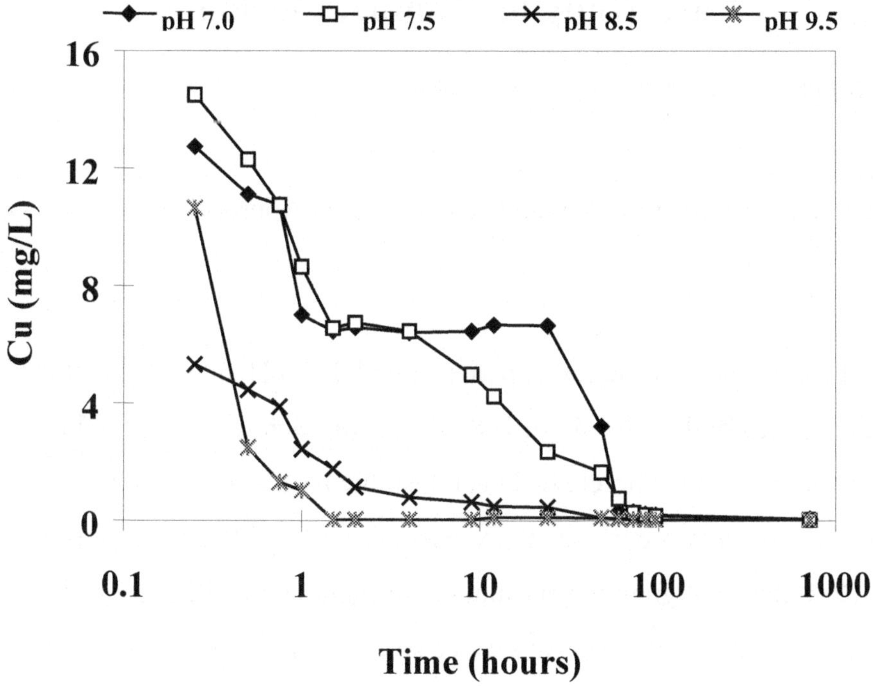

Figure 5.5. Effect of pH on the aging of $Cu(OH)_2$ to CuO. Aging is much faster at higher pH.

Case Study 6: Tampa, Florida

Host and Co-author: John Pierpont, Chief Operator

Key Experiences:
Low pH Coagulation and Lining of Concrete Using Different Approaches

Background:

The City of Tampa has substantial experience with low pH alum coagulation using sedimentation basins built in 1960. In 1983 the basins were inspected and it became evident that about 2" of concrete had been lost, the basins were leaking significantly, and some rebar was completely exposed. To rehabilitate the basins they were sandblasted, 2-4 inches of gunnite was applied throughout and then surfaces were coated with epoxy. Another set of basins in service since 1974 had the same problems, but rehabilitation required only sandblasting and application of the epoxy since losses of concrete were less significant. The epoxy has performed extremely well in the opinion of the plant personnel. Since the rehabilitation the plant changed to a combination of ferric and sulfuric acid coagulation at pH 4.0-4.2—no real problems have been experienced since.

Figure 6.1. Sedimentation basin lined with epoxy in 1983 was drained for inspection. Close viewing of epoxy coating illustrated it was still in good shape in 2000 after many years of coagulation at pHs as low as 4.0 using iron and sulfuric acid (upper right), and metallic components in the basin had also been coated and were still in good shape although general corrosion was apparent (lower right).

Metallic components in the plant have had very few problems with the exception of cast iron and steel impellers on some pumps that were failing every 2 to 6 months. This problem has been solved by gradual replacement of parts with epoxy or ceramic coatings, or with stainless steel parts which have never failed. In addition, some discharge pipe from the sludge thickener down to sludge processing failed after about 14 years of use (Figure 6.2), but such failures are not unexpected. Most recently the plant attempted to fix a problem from concrete corrosion in their flash mix basin. This basin had been unlined from about 1960 until the 1980's when it was lined with epoxy. Inspection in early 2000 revealed holes between 4-8" deep in the 1 foot thick wall and rebar was exposed. The holes were right next to the points of acid and coagulant addition, therefore, pH of the water at this point was very low and mixing rate was also high. The concrete was sandblasted, all holes were patched with hydraulic cement, and then all concrete surfaces were then coated with a rubberized material (Figure 6.2). Portions of this coating were detached and were floating on the water surface during the site vist-- the usefulness of this approximately $150K rehabilitation project was still being evaluated as of December 2000.

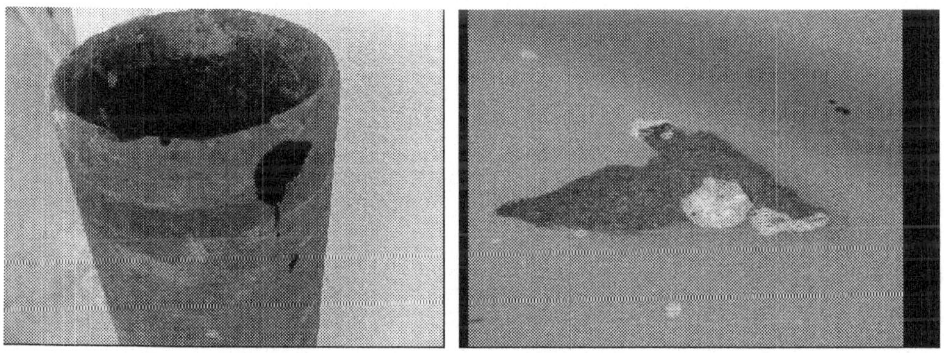

Figure 6.2. Picture of failed pipe section transporting solids from thickener to solids handling (left), and rubberized protective coating that had detached from concrete and was floating on water surface (right).

The plant started feeding 93% sulfuric acid in 1990 to enhance coagulation. Carbon steel was originally used to pipe the acid, but this completely failed after 2 years and they replaced it once using pipes lined with teflon. They also lost 2 36" butterfly valves where the H_2SO_4 mixed into the pipe. The plant has had very good luck with HASTALOY-C or Alloy 20 metal in applications exposed to the 93% acid.

Figure 6.3. Some unprotected components, such as the edge of this weir, had only lost about 3/8" of concrete due the low pH water since construction in 1986.

Case Study 7: Raleigh, NC

Host and Co-author: Larry McMillian

Key Experiences:
Concrete loss during enhanced coagulation, epoxy lining of sedimentation basins

Background:

The City of Raleigh, North Carolina switched from alum to ferric chloride coagulation in 1997, initially to improve the quality of their sludge for land application, but obtaining increased TOC removal from using ferric has now become increasingly important. When using alum coagulation pH was 5.5-7.0, but was decreased to 4.8-5.1 using ferric sulfate at a typical dose of 50 mg/L at this 86 MGD plant. During the site visit staff were queried closely about failure rates of pipe, pumps and other metallic components from corrosion. The consensus opinion is that the rate of problems has not increased significantly. The plant is gradually replacing certain components to more corrosion resistant stainless steel as part of a general upgrade program.

The plant has had significant problems with their sedimentation basins originally constructed in 1966 and in operation for 36 years at the time of this writing. Significant cracks have appeared on the bottom of the basin leading to significant water losses (Figure 7.1 and Figure 7.2). While nothing has been scientifically documented, it is generally believed that cracking and the rate of concrete degradation increased markedly when ferric use began. On the other hand, some of the newer concrete that was in use since 1983 looks fine, so it is possible that the concrete in the sedimentation basin is just showing signs of aging.

To remediate this problem, on the date of the plant visit, contractors had begun repairs that included cleaning and drying of cement (Figure 7.1), application of Quickrete hydraulic water stop or SikaRepair 224 sprayable mortar or Sikadur 31 Hi-Mod gel structural epoxy to fill cracks (Figure 7.1). Thereafter, the entire basin was to be lined with epoxy. It was anticipated that these changes would arrest corrosion of rebar within the basin, a factor that was just becoming noticeable and which prompted the repairs.

Another factor noted by plant personnel was that rust colored ferric coatings on the concrete and plant equipment take some getting used to. Given that training of operators stresses hygiene, the initial inclination was to wash all exposed stained surfaces frequently with high pressure hoses; however, it is quite likely that these high pressure washes were removing protective coatings from the concrete and adding to degradation from a physical abrasion rather than chemical attack. Interestingly, similar use of pressure wash in Ottowa and Buffalo Pound was discontinued due to suspected degradation of concrete even when alum coagulants were used. It is easy to understand an initial desire to increase washing due to the darker color of ferric staining, and it is also noteworthy that ferric staining would tend to hide early signs of rebar corrosion.

Figure 7.1. Workmen actively fill holes and cracks in concrete within the sedimentation basin (left and lower right) and dry concrete with heat (upper right) to prepare concrete.

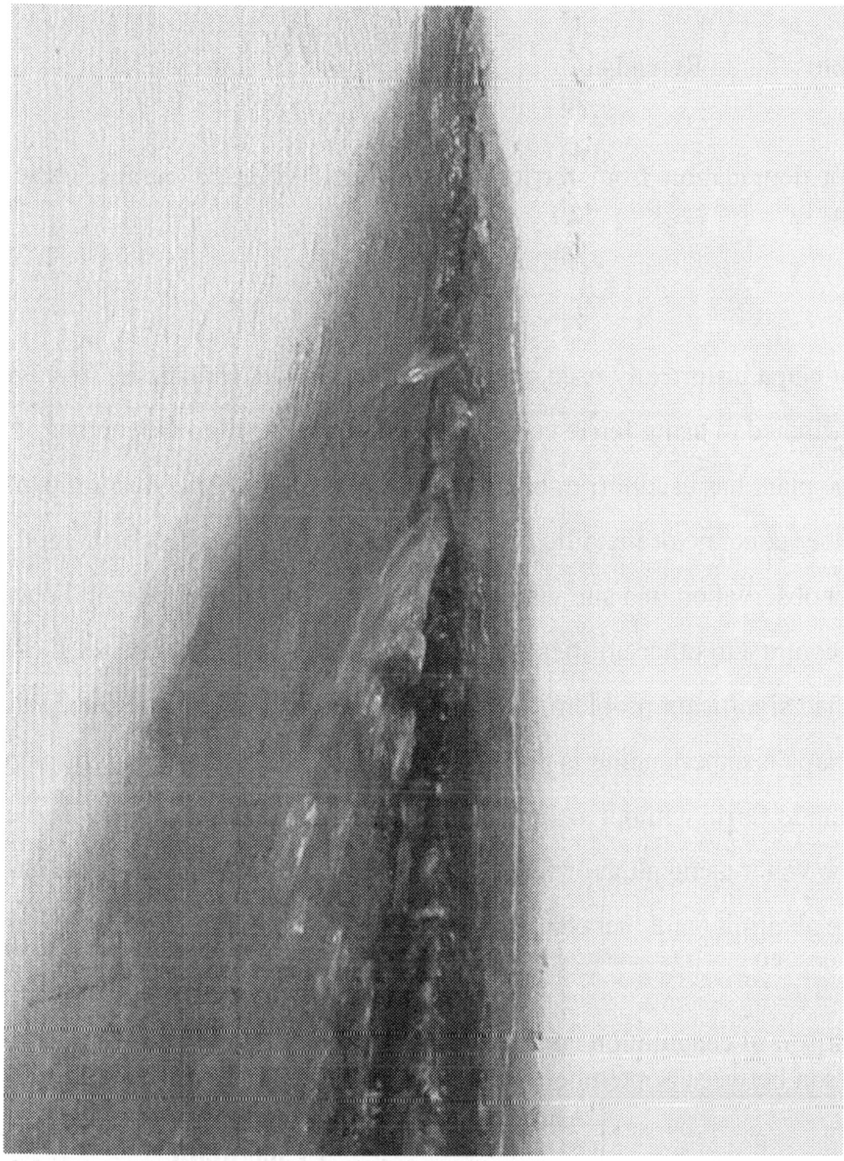

Figure 7.2. Most cracks in concrete leak below ground and are only noticed in water loss calculation or in flow that appears around plant. Occasionally leaks are visible on the side of a wall above ground, giving an idea of the magnitude of the problem that is typically out of sight.

Case Study 8: Napa, CA

Host and Co-author: Turan Ramadan

Key Experience:
Lack of significant degradation from ferric, clogged ferric chloride pumps, algae problem in sedimination basin

Background.

The City of Napa historically used only alum at their Lake Hennessey treatment plant, but in 1999 they switched to using ferric coagulant from about April to November of each year. In the two years the plant has used ferric, plant personnel have not noticed significant deterioration of either concrete or metallic components in the plant. This conclusion was supported by direct observation in a site visit by the AWWARF project team in February 2002.

In marked contrast to other utilities visited in this report, which were selected in large part because they had significant problems with infrastructure from enhanced coagulation, the lack of problems Napa is experiencing is probably typical of the vast majority of plants switching from alum to ferric without also markedly reducing coagulation pH. Typical analytical results on water using alum or ferric reveals the water has considerable magnesium hardness, moderate alkalinity and that coagulation pH rarely drops below 6.7.

Table 8.1. Comparison of coagulation conditions before and after switching coagulant chemicals (based on comparison between September 2000 and October 2001.

Parameter	**Alum Coagulant**	**Ferric Coagulant**
Date Constructed	1982	
Typical Flow (MGD)	20	
Coagulation Specifics		
Coagulation pH	7.35	6.95
Coagulant Dose (mg/L)	26	26
Alkalinity (mg/L as $CaCO_3$)	129	81
Calcium Hardness (mg/L as $CaCO_3$)	49	
Total Hardness (mg/L as $CaCO_3$)	146	

The plant did have two other interesting observations to report which are connected with goals of this study. First, a major problem the plant experiences is excessive algae growth in sedminentation basins (Figure 8.1). Not only does algae growth contribute to taste and odor problems, but on a few occasions so much O_2 was produced that tube settlers were lifted high out of the water, causing them to break when they settled back down. Because the clarity of the water in the sedimentation basin is so much higher using ferric than alum, light is able to penetrate deeper into the sedimentation basins fueling even more algae growth. Every few weeks the plant had to be shut down to pressure wash the algae biomass from surfaces, a process that could lead to physical degradation of the concrete as noted earlier. The solution to the problem, innovated by Plant Superintendent Turan Ramadan, was to install pool covers over the sedimentation basin. This dramatically reduced algae growth and the surprising range of problems associated with their activity.

A second factor noticed by plant personnel is that the solids content of ferric chloride chemical varies dramatically. Chemical from one supplier clogged the chemical pumps and pipes at the plant within just a few days, requiring cleaning to return pumping efficiency. The solution was to install a filter before and after chemical flowed into the holding tank. Every few weeks the filter bag is replaced and problems from ferric on pumps and pipes have disappeared. Since the plant changed suppliers, the new ferric coagulant has been essentially particle free, and while the filters are still in place they have not needed to be changed.

Figure 8.1. A key problem at the treatment plant is growth of algae in the sedimentation basin (upper left). On some occasions bubble from photosynthesis lifted tube settlers high out of the water, cracking them when they settled back into position (upper right). The solution was to install pool covers over the sedimentation basin to block the sunlight (lower right).

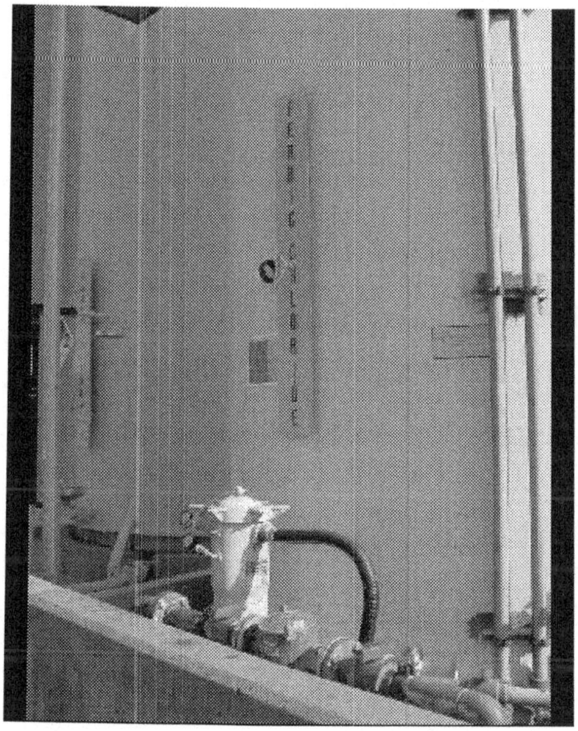

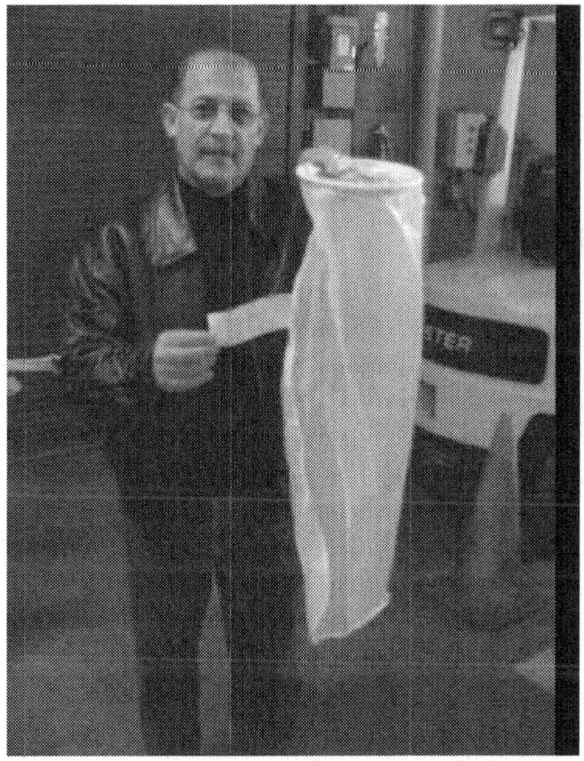

Figure 8.2. Photo of pre-filter installed outside of ferric chloride tanks (left) and liner of filter held by Turan Ramadan (right). So much particulate matter was in ferric chemical from some suppliers that operation of pumps and pipes were impacted.

Case Study 9: San Diego, California

Host and Co-author: Ernesto Fernandez and Steven A. Williams

Key Experiences:
Concrete loss during enhanced coagulation, epoxy lining of sedimentation basins

Background:

Infrastructure at the City of San Diego Otay filtration plant was profoundly impacted by enhanced coagulation. Fortunately, the city was well prepared to detect and remediate this type of problem in that they have a group of 8 full time employees dedicated to controlling aspects of corrosion in water and wastewater treatment.

The concrete basins at Otay were installed in 1988 and were inspected annually by the corrosion personnel. The concrete was ASTI C-150 Type II Portland cement with Pozzolith, air entraining admixtures and surface treatment. From 1988-1994 the ferric chloride dose at the plant averaged 14 ppm with a typical coagulation pH of 7.2 in the rapid mix basins. In August 1994 the ferric chloride dose was increased to 30 ppm with typical coagulation pH 6.6. Inspections in October 1994 revealed obvious deterioration of concrete in the flocculation and sedimentation basins, filter beds, settled water conduits and sludge handling. Most prominently, piles of sand up to 4" deep were noticed in the rapid mix basin for the first time, and given the obvious friable nature of the concrete surface exposed to the water, it was suggested the sand originated from the concrete. This prompted a detailed in-house study to quantify the rate of attack using hammer and abrasion tests at selected locations in the plant.

Measurements in December 1994 (Figure 9.1) and then again in March 1995 indicated that attack rates below the water line were as high as 0.210 inches per year, with an average of 0.072 and 0.139 inches per year in parallel sedimentation basins 1 and 2, respectively. Aggregate was exposed in many locations, whereas concrete above the water line was not softened or deteriorated significantly from the original construction (Figure 9.1). Measurements of water quality at the rapid mix indicated a typical pH of 6.63, alkalinity of 48 mg/L as $CaCO_3$, calcium hardness of 90 mg/L as $CaCO_3$, 140 Cl^- and 95 mg/L SO_4^{-2}. It appears that the sample location in the rapid mix might basin have had more than a typical dose of $FeCl_3$, since just before sedimentation pH had increased to 7.1, alkalinity increased to 80 mg/L as $CaCO_3$ and Cl^- had decreased to 114 mg/L while other water quality parameters remained the same. This highlights a possible problem in that certain areas of concrete may be exposed to higher than normal concentrations of coagulant before mixing is complete.

In October 1996, the plant was inspected by PSG Corrosion Engineering to examine the quality of concrete at the Otay Plant. That inspection determined the existing concrete was suited to a protective barrier coating if the surface was properly prepared and a new layer of polymer-modified Portland cement was installed. The estimated total repair cost of $852-1,231K included $372-429K to replace the filter media and $264-430K repair for just the flocculation and sedimentation basins. Destructive testing of the concrete also revealed that the overall quality of the concrete used in construction was fair to poor based on an estimated water cement ratio of 0.50-0.55 during placement. Chloride had penetrated between 1/4-3/4" with the highest chloride in concrete from the flocculation basin. Samples throughout the plant were carbonated to a depth between 1/8-1/4." Thankfully, the chloride had not yet penetrated to the reinforcing steel throughout the plant, given that the average depth of concrete coverage over the steel was more than 2" as revealed by a Soiltest Microcover Magnetic Meter. However, a few areas were noted where concrete coverage was only 0.25" and these were repaired using SikaTop 123 Plus polymer-modified mortar.

The PSG report then discussed different repair options (Table 9.1). They initially recommended a urethane system due to its elastomeric properties and relatively low cost compared to a new coating of silica fume cement. However, the system actually selected by San Diego was Hydro-Pox two component, VOC free 100% solids poly-amine epoxy coating. This material can be applied at a variety of temperatures, has high chemical resistance and only requires 3 days to be fully cured. The complete installed cost for the surface preparation, primer, and then a thin film top coat of 16-30 mils epoxy was about $\$7/ft^2$. The total installation time was approximately 90-120 days after staging treatment for each of the two parallel basins and the total area ultimately treated was over 100,000 ft^2.

Table 9.1. Summary of repair options outlined in PSG report and input from Steve Williams. All barriers require abrasion blast or high pressure pre-cleaning to remove loose surface materials. The approaches using epoxy or polyurethance also require repair of concrete at large voids and then smoothing the surfaces with leveling mortar.

	Type of Protective Coating		
Parameter	Silica fume coating	Polymer modified Portland cement and 100% epoxy	Polymer modified Portland Cement and 100% elastomeric polyurethane coating
Depth of Coating	1" deep	16-30 mil	60 mil
Estimated life	10-12 years	10-13 years with minor repair at 5 years	10-13 years with minor repair at 5 years
Preliminary Estimated Cost ($\$/ft^2$)*	$15-20	$about the same as polyurethane	$5.60-9.12*
Proven performance	Relatively new without long term case history	> 40 years experience	> 10 years experience, but rarely used for potable water
Other factors		Little resistance to concrete movement	Very fast setting, somewhat resistance to concrete movement

*Range based on estimates from three coatings contractors and 85,600 square feet total area.

The site visit by the project team was timed to coincide with a major inspection of the coating planned 3 years after installation (Figure 9.2). The sedimentation basin were drained, revealing that the epoxy was in very good shape overall. Minor blisters were noted at a few locations, and a few blisters were popped to sample the solution underneath (Figure 9.2).

In general, most metallic components at this plant were in good shape, perhaps because they were protected by Mg anodes (Figure 9.3). Another type of infrastructure degradation has been noted at the Miramar Treatment plant where concrete degradation was noticed in a collection conduit (Figure 9.4). This conduit does not run completely full, and substantial spalling of concrete and rebar corrosion is occurring along the top just downstream of a Cl_2 injection point while very little deterioration is occurring upstream of the injectors. Ernesto Fernandez has hypothesized that the mode of degradation is direct attack of degassed Cl_2 on the concrete, which reacts to form HCl.

This reaction also causes a very high concentration of Cl^- on the surface that accelerates rebar attack. Staff are monitoring degradation at this location, which is difficult to repair because the pipe cannot be readily taken out of service.

Figure 9.1. Initial inspection of basin below waterline and above waterline indicated significant softening of concrete from enhanced coagulation (left). Concrete surfaces were scraped (white area on right) to measure depth of degradation at various points in the basin.

Figure 9.2. Picture of sedimentation basin at San Diego during inspection of epoxy coating 3 years after installation (basin). The coating was in good overall shape, with a few blisters (lower left) and very few breaks in the coating (lower right).

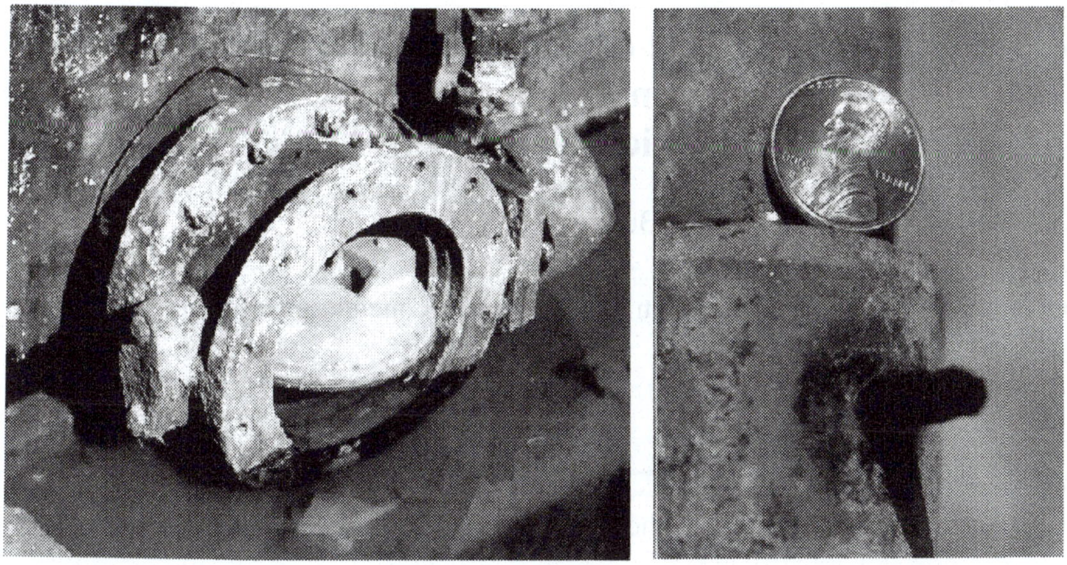

Figure 9.3. Metallic components at San Diego were in good shape but most had been protected using Mg anodes (left). Some small, unprotected metallic parts like this 304 stainless steel pin have failed, which seems trivial but actually causes problematic failure of a system used to open a valve (right).

Figure 9.4. Concrete degradation and spalling from top of water conduit exposed to Cl_2 gas.

Case Study 10: Direct Test of Concrete Corrosion During Coagulation in Potomac Water Using PACl, Ferric Chloride and Alum

Hosts and Co-authors: Joe Yorke and Bob Buglass

Key Experience: A test on concrete corrosion as a function of concrete specifications, Coagulation pH and Coagulant Type

Background:

One key question in this research is whether alum, ferric or PACl coagulant is inherently more corrosive to concrete if all other factors are equal. Surprisingly, very little work has been conducted on this subject. At the time this project was underway, the City of Fairfax, VA was building a brand new treatment plant. Fairfax has long used a combination of alum/PACl coagulants in treating Potomac water, but was considering use of ferric chloride at the new facility. In addition, the Washington Suburban Sanitary Council (WSSC) draws water from roughly the same spot in the Potomac River as Fairfax (Table 10.1), but uses exclusively PACl coagulant. The City of Fairfax partly funded a head to head comparison of concrete corrosion in the presence of alum, ferric and PACl coagulants in this raw water.

Table 10.1. Typical Raw Water Quality of Potomac River Source water

Parameter	**Value**
Raw Water Color (mg Pt/l)	15-40
Average Turbidity (FNU)*	8-180
Raw Water pH	7.8-8.4
Raw Water Alkalinity	45-90
Silica (mg/L as Si)	1-3.4
Sulfate (mg/L)	15-50
Chloride (mg/L)	20-60
TDS (mg/L)	130-225
Alum and Ferric Exposure* Dates: 4/25/01-2/15/02 or 9.6 months	Max pH = 6.9 Avg pH = 6.4 Min pH = 6.0
PACl Exposure Conditions Dates: 7/27/01-2/16/02 or 6.6 months	Max pH = 8.0 Avg pH = 7.8 Min pH = 7.6

*Ferric exposure pH was approximately equal to that of alum > 90% of the time, but due to infrequent failures of the feed apparatus, pH was occasionally equal to that of the raw water.

The test proceeded roughly as follows. First, two sets of concrete coupons were poured at Virginia Tech, one of which met specifications for concrete used at the new Fairfax facility and a second set with air content exceeding those specifications. Coupons were enclosed in mesh (Figure 10.1) and placed either 1) in the rapid mix basin at Fairfax using alum coagulant, 2) in a 55 gallon drum rapidly mixed, to which ferric chloride was dosed upstream to maintain the same coagulation pH as for those coupons exposed to alum (Figure 10.2), or 3) in the rapid mix basin at the WSSC Potomac treatment plant using PACl and a higher coagulation pH (Table 10.1). At the end of the test, the coupons were subjected to a variety of tests designed to examine corrosivity. Results are presented after providing details of the Materials and Methods.

Materials and Methods:

Preparation of Coupons.

A concrete mixture was proportioned for a one-half inch maximum coarse aggregate. The mixture was proportioned to meet the Fairfax County Water Treatment Authority Project 01800-6, Griffith Water Treatment Plant, Specification Section 03300, CAST-IN-PLACE CONCRETE. The one-half inch maximum coarse aggregate size was selected in order that the distance between the sides of the 2-inch cube molds was at least three times the maximum nominal aggregate size. Standard 2-inch cube specimens were selected to maximize the accuracy of subsequent concrete deterioration measurements as weight loss and chloride and other ion ingress.

The concrete ingredient properties and characteristics for the aggregates, Portland cement, and water and admixtures are presented in Appendix Tables 1-3A, respectively. With the exception of the Portland cement alkali content, the materials meet project specifications. The Portland cement alkali content was 0.72 percent which exceeds the specified low alkali cement maximum allowable content of 0.60 percent. However, both the natural sand and crushed siliceous gravel are from sources that have no field history of being alkali reactive.

The mixture was proportioned in accordance with American Concrete Institute (ACI) practice (Appendix Table 4A). At first blush it would appear that the ACI recommended air content for severe exposure of 6 to 9% is somewhat greater than the project specification of 5 to 7%. However, the project specification is for the total volumetric air content of the concrete after placement. Whereas, the ACI recommended air content is normally based on air content measurements taken prior to placement. As it is normal to lose a portion of the total air (entrapped plus entrained air) during the placing and consolidation of the concrete, the air content prior to placement needs to be somewhat higher than the specified air content after placement.

Two concrete batches were prepared and a total of 54 two inch cubes were cast from each batch. The second batch was made because the slump and air content of the first trial batch was greater than project specifications. The second batch meets project specifications, except for the air content. However, as previously stated the project specification for air content is for after placement and the measured batch air content is to be considered as before placement. It is our opinion that the in-place air concrete of Batch 2 will be within the project specification limits and thus Batch 2 meets the project specifications.

To simulate field curing conditions, the concrete cube specimens were cured in saturated lime water at 70 +/- 2°F for 7 days and then placed in a controlled environment of 70 +/- 2°F and a relative humidity (Rh) of 50 +/- 3% for a minimum of 30 days prior to removal and subsequent exposure to water treatment plant environments. After removal from the exposure environments and prior to condition assessment, the cube specimens shall be returned to the controlled environment of 70 +/- 2°F and 50 +/- 3% Rh for the same preconditioning time period. During the time period from after removal from the controlled environment and immediately prior to the exposure to the water treatment plant environments, the cubes will be kept in a container with a Rh of 50 +/- 3%.

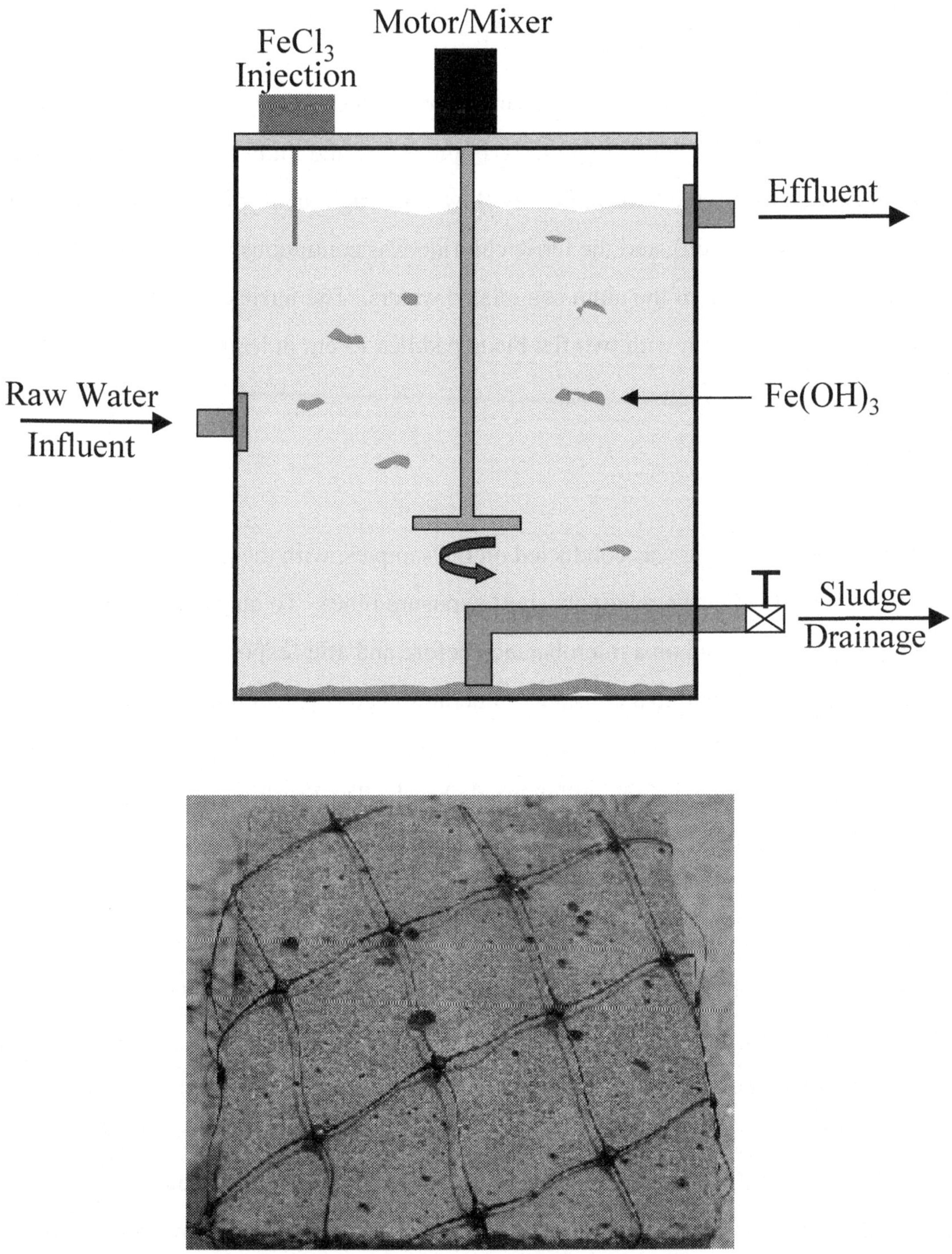

Figure 10.1. Fifty-five gallon exposure apparatus for the ferric rig (above) and concrete cube within mesh holder (below).

Coupon Exposure.

For each type of coagulant tested, five samples of each concrete coupon type (10 samples total) were attached to a PVC holder using mesh (Figure 10.1) and immersed the test waters after coagulation. Raw water from the Fairfax facility flowed through the ferric chloride exposure rig at 1.3 gallons/minute (4.9 L/min), and the ferric chloride was continuously added such that the final solution pH matched that of the alum coagulated waters. The ferric chloride solutions were continuously mixed at 1725 rpm with two flat blade paddles 15 cm in length.

Analysis of Coupons.

A wide variety of testing was conducted on the samples, with the goal of detecting major differences in corrosivity after the relatively short exposure times. To quantify weight loss, the prepared samples were weighed on a microbalance before and after exposure. Prior to each weighing, the samples were held in a constant temperature and constant humidity room for one month. After the exposure, the samples were weighed without any cleaning and after 10 even strokes on the face of each cube using a nylon bristle brush. The brushing removed much of the attached dirt and loose aggregate. Weeks of effort attempted to quantify corrosivity using SEM and EDAX, Cl^- penetration testing, and carbonation depth profiling. SEM and EDAX proved to be unsuccessful since a vacuum could not be maintained on the porous concrete samples. Cl^- penetration (Otsuki, 1992; Hooten, 2001) uses a colorimetric dye, and in our samples at least, the dye did not develop detectable color, perhaps because Cl^- penetration into the sample was insignificant. Likewise, the depth of carbonation in the concrete (Figure 10.2) was not detectably different between samples after the short-term exposures tested (Broomfield, 2000). This left weight loss as the primary measure of water aggressiveness to concrete. It is noteworthy that sample weight loss is generally considered the most significant measure of corrosivity, if statistically meaningful differences between samples are present (AWWARF, 1996).

Figure 10.2. Normal section cut through a concrete coupon (right), versus a coupon after cutting and phenolphthalein staining (left). Areas neat the surface that are not stained pink have had the lime neutralized by acid, whereas those stained pink have a great deal of active lime.

Results and Discussion

Visual examination of the coupons after exposure (Figure 10.3) indicated only minor surface degradation for coupons exposed to PACl and ferric, whereas major degradation was obvious for coupons exposed to alum. Although there was a surface coating on the coupons exposed to alum somewhat reminiscent of that observed at the Regina treatment plant, it did not appear to be protective of the concrete relative to the other coagulants.

The weight loss data provided a dramatic confirmation of the visual trends. For coupons exposed to PACl, a very small weight gain was observed, most likely due to new deposits and precipitates such as floc sticking to the surface after exposure (Figure 10.4). Most of these deposits were removed by cleaning, and as a result virtually no weight change was observed during the 6.6 month exposure to PACl at the WSSC Potomac treatment plant. The coupons exposed to ferric were noticeably stained, but less than 1.7% weight loss occurred after a 9.6 month exposure at Fairfax. In contrast, 3.4% of the total coupon weight was lost before cleaning for the alum exposure, and an additional 2.3% of the weight was lost after cleaning for a total weight loss of 5.7%.

Comparing Type 1 and Type II concrete indicates that, in general, the batch with excessive air lost more weight than did the batch without excessive air. This trend was sometimes, but not always, significant at 95% confidence. Thus, it is clear that concrete type plays a clear role in the extent of degradation. Overall, given the high specifications for the concrete to be used at Fairfax, a relatively large amount of degradation was observed. However, to say the least, it does not appear that ferric is more aggressive to the concrete than the alum, since the relative ranking was weight loss increasing in the order PACl>Ferric>Alum.

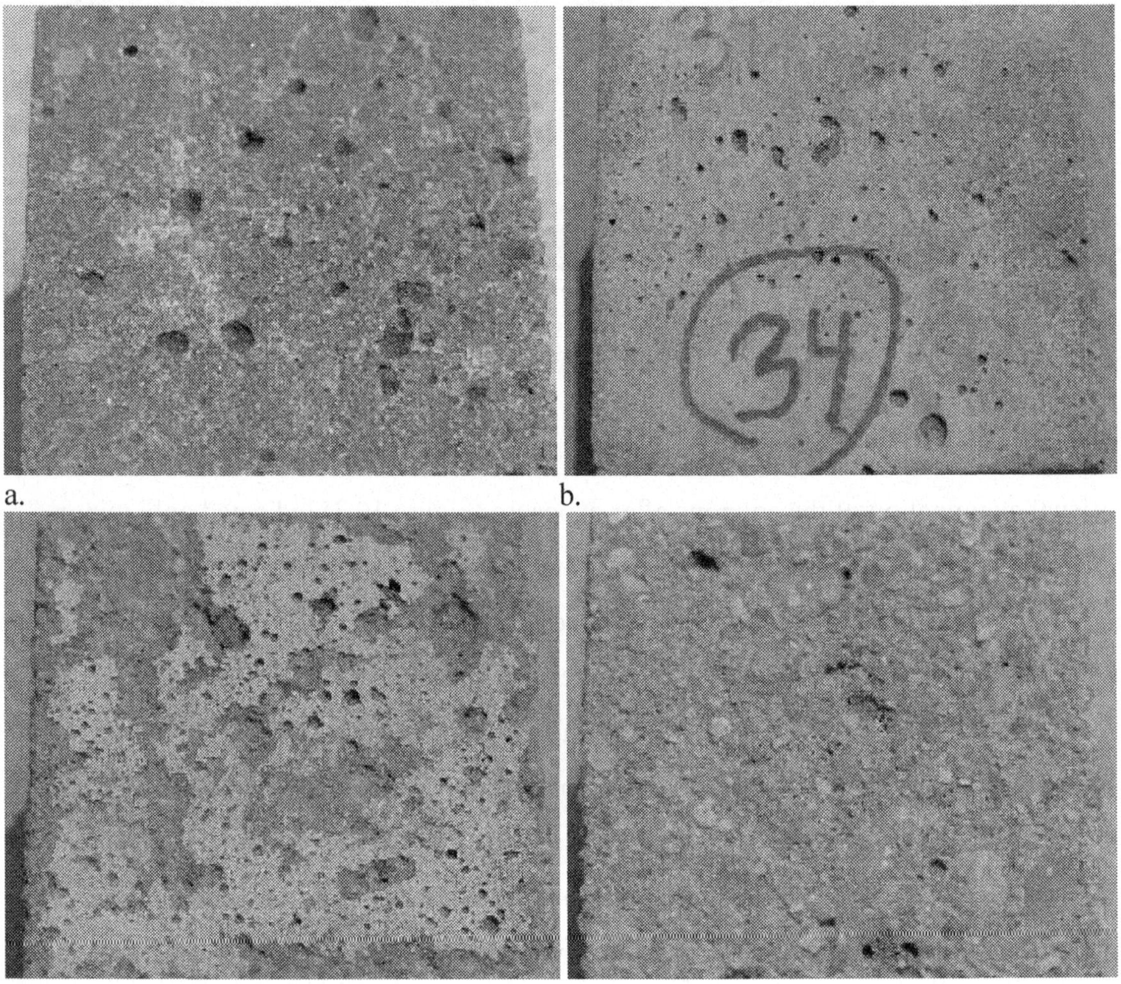

Figure 10.3. Final Concrete Photos – a. Ferric Chloride, b. PACl, c. Alum before final cleaning, d. Alum after final cleaning

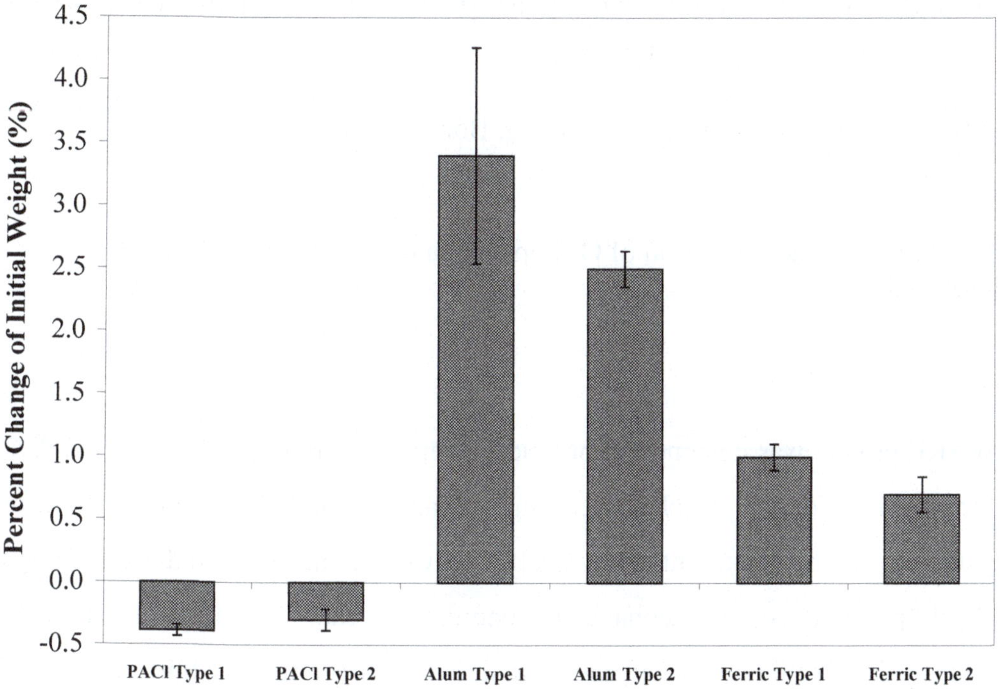

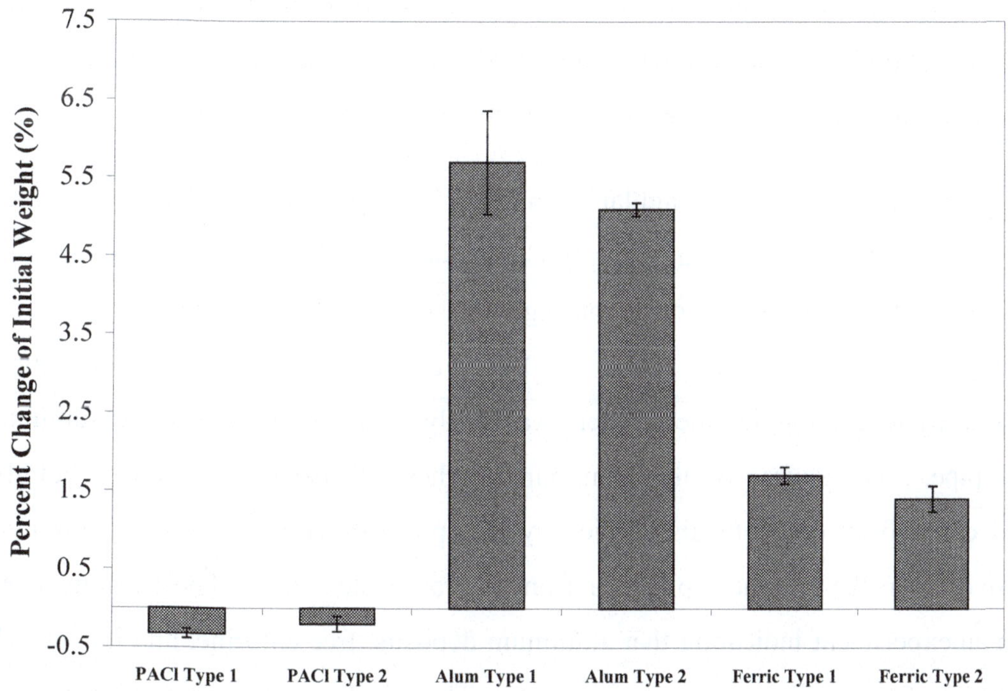

Figure 10.4. Percent weight change after exposure to the indicated coagulant before cleaning (above) and after cleaning (below). Type 1 refers to specified concrete with excessive air, whereas Type 2 meets all Fairfax specifications. Graphs indicate 95% confidence interval on 3 or 4 samples per condition. Positive values indicate a weight loss, and negative values indicate a weight gain.

Case Study 11: Potential Role of Orthophosphate in Inhibition of Copper Pitting

Hosts and Co-authors: Becki Marshall, Jason Rushing, Bob Buglass

Key experience:
Natural Inhibitors to Pinhole Leaks, Addition of Orthophosphate to Supplement Natural Inhibition of Pinhole Leaks

Background:

This AWWARF project was designed to produce several important products for utilities. All of these efforts were directed at characterizing impacts of enhanced coagulation on infrastructure degradation at the treatment plant itself. However, the case study at Billings, Montana highlighted predictions that enhanced coagulation can, under at least some circumstances, increase the likelihood of pitting corrosion of copper tube within consumers' homes (Edwards et al., 1994a). That is, after changing from alum to ferric, this utility started experiencing numerous pinhole leaks in copper tubes within their treatment plant. The evidence that a change in coagulant caused the leaks is circumstantial but quite strong.

Another case where changes in coagulation might have impacted pinhole leak frequency was reported by the Washington Suburban Sanitary Commision (WSSC), whose customers were experiencing an outbreak of cold water copper pitting corrosion in homes. In previous research work at this utility described in Rushing (2002 a; 2002 b) and Edwards (2002), circumstantial evidence was gathered suggesting that the problem was partly caused by formation of aluminum deposits on the pipe. The source of the aluminum in these deposits is not clear, but the aluminum could come from concrete distribution system pipes or concrete lining, aluminum coagulant carryover from the treatment plant, or from the raw water itself. The Rushing work culminated with an experiment indicating that aluminum deposits, free chlorine, and higher pHs could initiate pitting of copper in at least some waters. Though the work was conducted with water synthesized in the laboratory, the bulk salt concentrations were representative of those present at WSSC and elsewhere.

As noted by Rushing et al. (2002a; 2002b), certain trends in industry practice including lead and copper corrosion by-product control, cleaning and lining of main pipes with cement, and use of PACl coagulants are likely increasing the pH of water reaching consumers homes as well as the deposition of aluminum on pipes. In the past, higher pHs were almost always deemed beneficial for copper corrosion, but pitting corrosion tendencies were worsened as pH increased to 9.0 in the Rushing et al. work. The overall implication is that some changes in water treatment practice including enhanced coagulation, might be increasing the pitting propensity of distributed water as was predicted by Edwards et al., (1994a).

The notion that water chemistry might be the exclusive cause for pinhole leaks is controversial, to say the least. However, water quality modifications may be the most cost-effective approach for protecting extensive existing copper pipes. Given this fact, and the high stakes to impacted consumers, it was deemed useful to carefully examine recent outbreaks of pinhole leaks around the country. Special efforts were made to determine whether they are occurring in situations where:

1) pHs at the tap have been increased above 8.0, even temporarily after concrete installation, corrosion control, or mains disinfection practices;
2) aluminum deposits occur on pipes; and
3) relatively high concentrations of oxidants (disinfectants) have been present, and if so, what are the actual concentrations and type of disinfectants.

Assuming that current hypothesis of Rushing et al. is correct, and changes to treatment are increasing the pitting propensity of many waters around the U.S., it would also be highly desirable to determine if we can stop the problem once it has started. Dosing of orthophosphate is among the most promising cures that have been suggested, since orthophosphate dosing has successfully controlled pitting outbreaks in some circumstances (Edwards, 2002). Since the activity of orthophosphate may depend on pH, a special focus of this investigation is to determine whether orthophosphate would be effective at higher pHs that are common to distributed waters that have contacted concrete.

Thus, the goal of this case study is to examine orthophosphate as a means of preventing pinhole leaks.

MATERIALS AND METHODS

Potential Use of Orthophosphate to Stop Pitting

Two sets of experiments were conducted. The first set of experiments used water synthesized in the laboratory. The second set of experiments utilized synthesized water in an initial exposure phase, followed by use of water shipped to the Virginia Tech research laboratory from WSSC.

In the first set of experiments, the role of phosphate on copper corrosion using synthetic water was examined. 5/8" nominal size copper couplings (surface area = 20 cm^2) were purchased from a local hardware store, rinsed with pure water, and placed in Reiber flow-through polarization cells (Reiber et al., 1989). Additionally, two 12 inch long, 5/8" nominal copper pipes were attached to each recirculating system, one of which was cut in half horizontally and sealed with clear plexiglas to visually observe the progression of scale development. Sixteen liters of synthesized drinking water was circulated through the copper coupons and Reiber flow cells at a flow rate of 1 gpm. The water initially contained 34 mg/L alkalinity as HCO_3^-, 14 mg/L SO_4^{2-}, 20 mg/L Cl^-, and 17 mg/L Ca^{2+} added as reagent grade sodium or calcium salts to filter sterilized distilled-deionized water. pH was adjusted to target values by adding 1 M HCl or NaOH, and no attempt was made to maintain alkalinity.

A total of four water quality conditions were tested at constant pH, aluminum, and chlorine levels. Two concentrations of inhibitor, 1 mg/L as P and 0 mg/L as P, were examined in duplicate at two pH levels, 7.7 and 8.3 (8 cells total). The target chlorine and aluminum levels were both 5 mg/L. The $Al(OH)_3$ stock solution was created by raising the pH of a Al-$AlCl_3$ solution to 8.0 using NaOH, allowing the solids to settle, decanting the water, re-suspending the solids in fresh water, until the conductivity of the suspension was less than that of the experimental water. To account for possible deposition of aluminum solids within the experimental apparatus, aluminum solids were re-dosed once every 4 days after each water change. Thus, the actual aluminum concentration could have been as much as double the target levels, and they could have been lower dependent on the loss to the apparatus surface or pipe samples. Photographs were taken of the samples before and after the pH change using a digital camera for a visual analysis.

The pH was adjusted and Cl_2 residuals were replenished daily using bleach (6% NaOCl). In all experiments, free chlorine was measured using the DPD colorimetric test per standard method 4500-Cl G [8]. pH values were held to ± 0.2 units of the target value. The water was completely changed weekly and the test was ended on day 34 of the experiment. On day 22, the pH of four samples was increased to 9.0 or 10.0 as specified in later discussion to examine the effects of higher pH on phosphate effectiveness.

Corrosion potential (E_{corr}) and polarization resistance (R_P) measurements were collected hourly using a multiplexer connected to a personal computer, using a polarization scan rate of 0.5 mV/sec from –5 to +5 mV of the open circuit potential. Platinum counter and Ag-AgCl reference electrodes were checked and maintained regularly to insure accuracy.

In the second set of experiments, after initiating active pits within the samples, the effects of orthophosphate in reducing pitting in real WSSC water were tested at constant pH, aluminum, and chlorine concentrations. The pits were initiated on 24 copper coupons by circulating synthetic water containing 5 mg/L chlorine through the pipes in the presence of 5 mg/L aluminum at pH 9.2. On day 36, a conditioning process was started where real WSSC water, without the addition of chlorine or aluminum solids, was circulated through the pipes at pH 9.2. Eight of the copper couplings were then placed in Reiber flow-through polarization cells, with two of the remaining copper couplings attached to each recirculating system.

Six of the eight samples examined contained 1 mg/L of phosphate and were tested in duplicate at pH 8.3, 8.7, and 9.1. The remaining two samples were used as controls without phosphate at pH 8.3 and 9.1. During this phase of the exposure, all samples had a target chlorine concentration of 4 mg/L but no aluminum was added to WSSC water. pH was adjusted and Cl_2 residuals were replenished daily using the same methods previously discussed. The water was completely changed as indicated and tests were ended on day 74 of the experiment. At the end of the experiment, photographs were taken of the corroded copper surfaces using a digital camera.

RESULTS AND DISCUSSION

Effects of Orthophosphate and pH on Copper Corrosion in Synthetic Water

Chlorine Decay Kinetics. Chlorine can react with copper surfaces according to the following balanced general equation:

$$OCl^- + Cu\ (metal) \rightarrow CuO + Cl^- \qquad \text{Equation 11.1}$$

As a result of this reaction, free chlorine is converted to chloride, and the copper metal is oxidized or corroded. While copper can also be corroded by reacting with oxygen, and chlorine can decay spontaneously by other pathways, the rate of chlorine decay is sometimes related to the copper corrosion rates.

Chlorine decay rates were calculated by tracking the loss of chlorine after 24 hours and calculating the rate in terms of mass Cl_2 lost per unit pipe surface area, ($\mu g\ Cl_2$ lost/cm^2). If the percent increase in chlorine decay rate in the system without orthophosphate is calculated as follows:

$$\%increase = \frac{(R - R_p)}{R} * 100 \qquad \text{Equation 11.2}$$

Where: R= decay rate without phosphate, $\mu g/cm^2$
R_p= decay rate with phosphate, $\mu g/cm^2$

A 0% increase indicates that chlorine decay does not increase, whereas a 50% increase indicates that the presence of orthophosphate cut the rate of chlorine decay in half. The presence of orthophosphate generally led to a 34-53% reduction in chlorine decay rates in the systems tested (Table 11.1). Somewhat surprisingly, the rate of chlorine decay was not a strong function of pH. The net result is that higher chlorine residuals were present throughout the study in the system containing phosphate, which can be expected to make the water more aggressive in the context of corrosion (Figure 11.1 and 11.2). However, the results also suggest that phosphate was also inhibiting the corrosion reaction between free chlorine and the copper metal, and was therefore protecting the metal.

Corrosion Potential. Electrochemical data were collected for each pH condition in duplicate (Figure 11.3 and 11.4). On day 23, 550 hours into the experiment, the pH of one of the duplicate samples at pH 7.7 and 8.3 was increased to examine the effects of high pH values on the effectiveness of orthophosphate. The pH 7.7 samples were raised to pH 9.0 and the pH 8.3 samples were raised to pH 10.0. Corrosion potential (E_{corr}) was affected by both pH and the presence of phosphate in the system. At pH 7.7, there was not a large difference in E_{corr} values of samples with and without phosphate (Figure 11.3). This is probably due to the fact that pitting is not observed at this pH in the first place (Edwards, 2002). At pH 8.3, the samples containing phosphate consistently yielded lower E_{corr} values (Figure 11.4). In the cases where pH is raised above 9.0, E_{corr} spiked upwards 100's of mV even if phosphate was present. We interpret this to indicate that, while orthophosphate greatly reduces the pitting type activity at higher pH, it does not completely inhibit spikes in E_{corr}. On the basis of existing understanding and the very limited data collected in this short term test, the suggestion is that phosphates would greatly reduce pitting frequency at pH 8.3, would be less effective at pH 9.0, and would be quite effective at pH 10 when compared to the synthesized water at the same pH without phosphate.

Table 11.1. Chlorine Consumption in Synthetic Water.

pH	Chlorine Consumption (ug/cm^2)		% Difference
	With P	*Without P*	
7.7	48.1	82.2	42%
8.3	54.3	82.2	34%
9.1	48.1	91.5	47%
10	43.4	91.5	53%

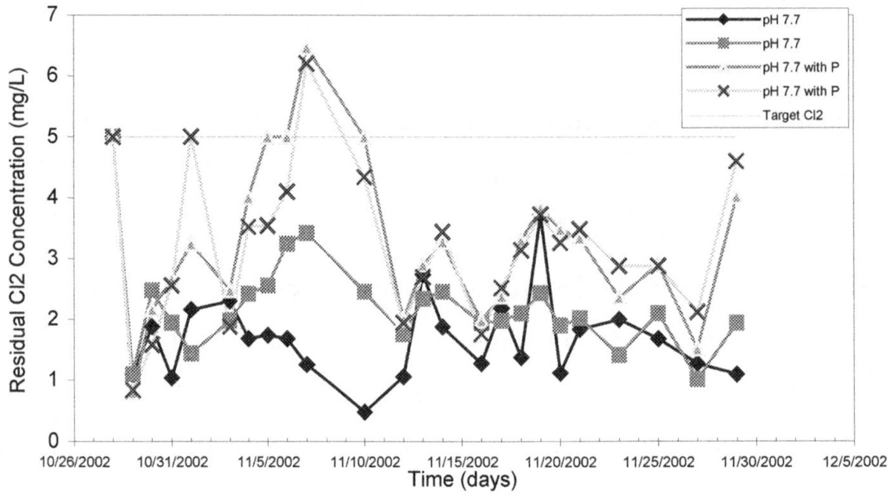

Figure 11.1. Daily Chlorine Residual at pH 7.7.

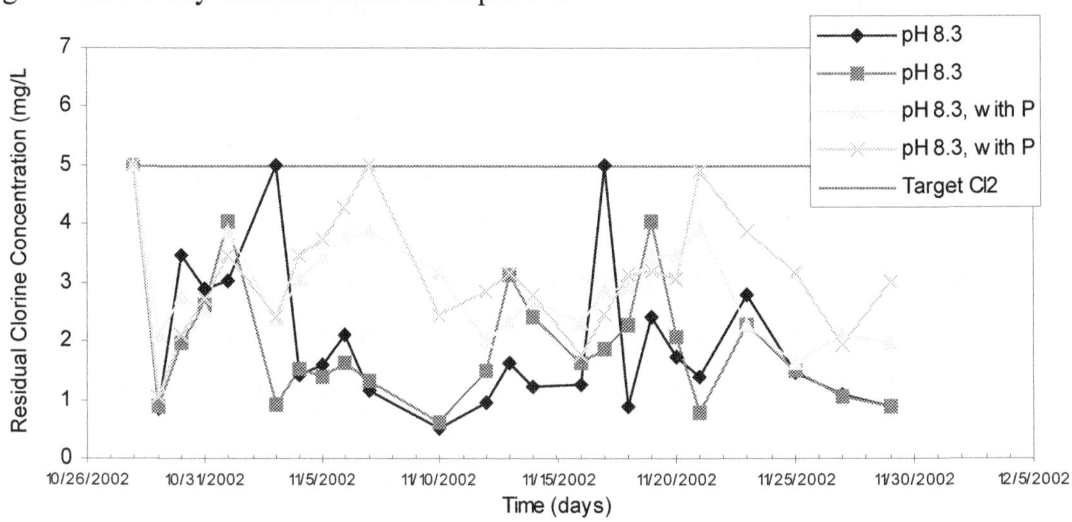

Figure 11.2. Daily Chlorine Residual at pH 8.3.

Another distinction was observed in the typical response to daily pH and chlorine adjustments (Figure 11.5). While samples containing phosphate momentarily decreased in E_{corr} when chlorine was re-dosed to target values and sodium hydroxide (NaOH) was used to increase pH, the samples without phosphate showed a rapid spike upwards in E_{corr}. This further demonstrates that orthophosphate is fundamentally changing the reaction between the copper metal and the chlorine, as was already established previously in Table 11.1. This is also interpreted to indicate that orthophosphate ameliorates the potentially harmful effects of chlorine on copper corrosion.

In general, at pH 8.3, the samples without phosphate had a larger daily fluctuation in E_{corr}. Specifically, E_{corr} typically varied by 40 mV in the system without phosphate, versus 13 mV with phosphate. This may be due to the greater variation of the free chlorine concentration within the water, since chlorine was decaying at a higher rate. It is not understood whether this is good or bad in thc context of pitting, although variations in E_{corr} (i.e., electrochemical noise) are often believe to be an indicator (albeit a poor indicator) of pitting.

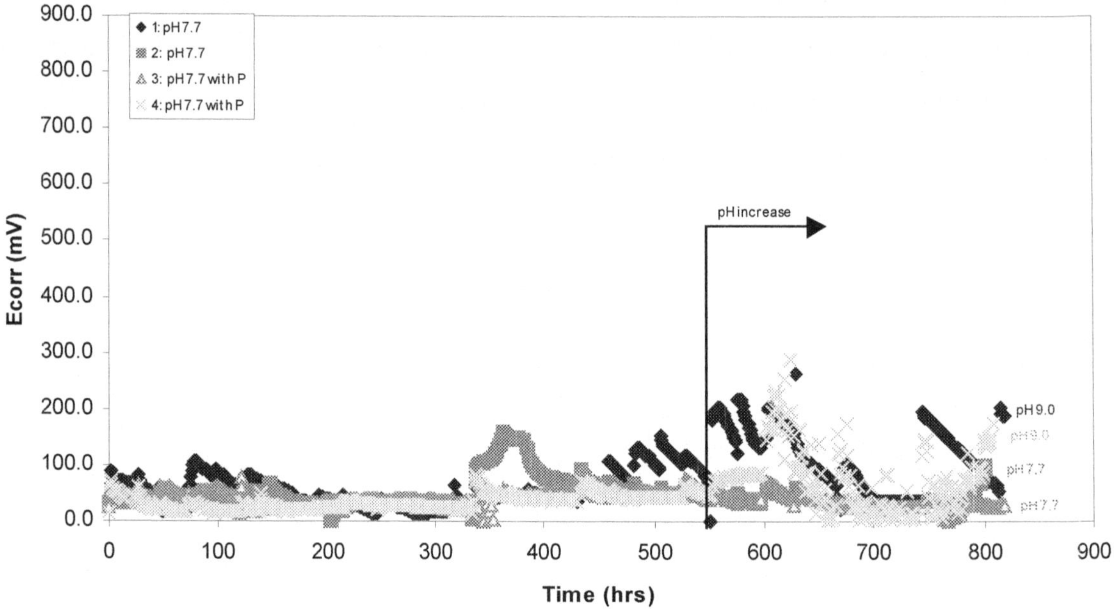

Figure 11.3. E_{corr} for copper exposed to synthetic water initially at pH 7.7, but then raised to 9.0 in a few cases.

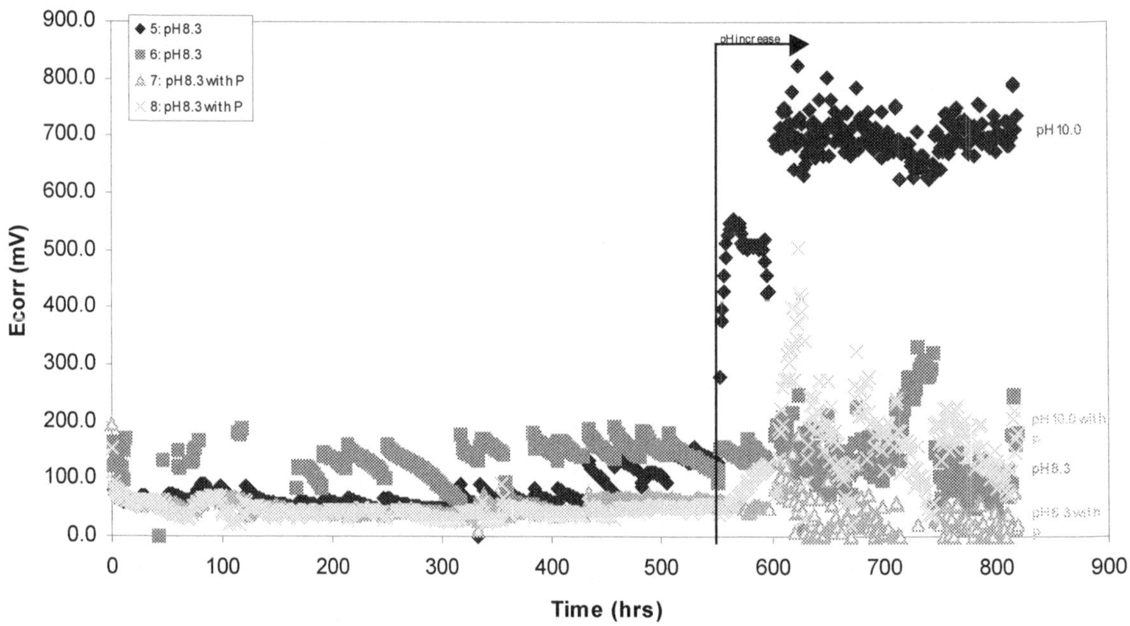

Figure 11.4. E_{corr} in synthetic water initially at pH 8.3, but then raised to pH 10.0 in a few samples.

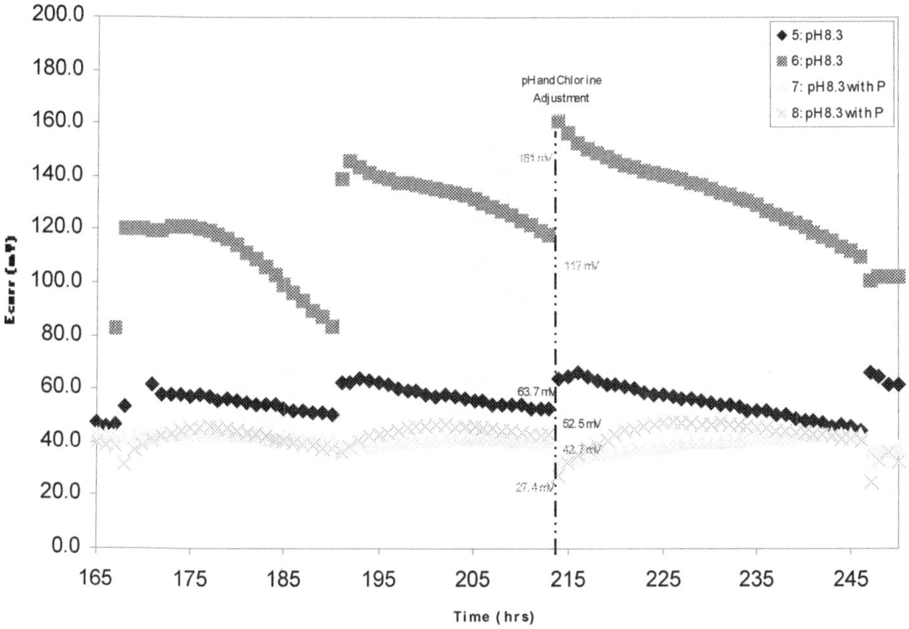

Figure 11.5. Typical E_{corr} Response to Daily pH and Cl_2 Adjustment.

The maximum corrosion potentials (E_{corr}) were determined for samples before and after the increase in pH (Figure 11.6). The selection of the maximum E_{corr} excluded data that was known to be faulty. To summarize the results, the E_{corr} was affected by pH as well as the presence of orthophosphate within the system. In all cases, the increase in pH increased the maximum E_{corr} attained. At pH 7.7, the sample with phosphate had a lower maximum E_{corr}.In contrast, when the pH was increased to 9.0, the sample with phosphate had a higher maximum E_{corr} than the sample without phosphate. At pH 8.3, the samples containing phosphate seemed to have an distinct advantage with lower maximum E_{corr} values. For instance, at pH 8.3, the maximum measured E_{corr} for the sample without phosphate was 156 mV compared to only 71 mV for the sample with phosphate at the same pH. When the pH was raised from 8.3 to 10.0, a large increase in E_{corr} was observed for samples with and without phosphate. Even though the phosphate appeared to have lost some of its useful effects at this much higher pH, the maximum E_{corr} was still much less when phosphate was present.

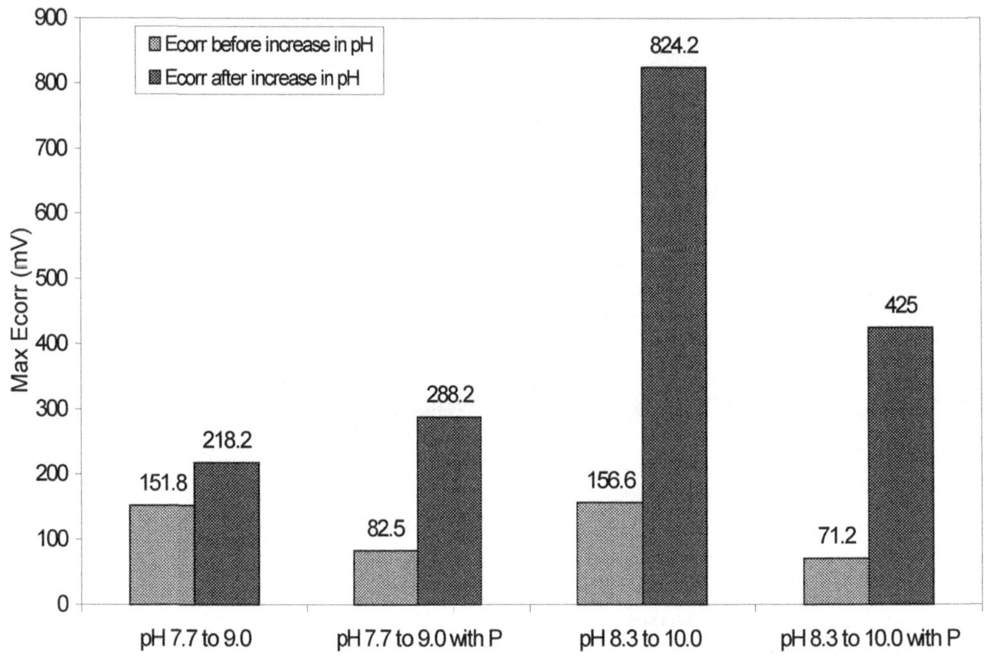

Figure 11.6. Maximum E_{corr} Readings Before and After pH Increase.

By comparing the samples at pH 7.7 and 8.3 without P, the effects of pH are obtained. The maximum E_{corr} readings for both samples were comparable, however, at pH 8.3 the E_{corr} values were consistently in the range of 250 mV, while the E_{corr} values at pH 7.7 were anywhere from 50 to 220 mV (Figure 11.3 and 11.4). The increase in pH had a much more profound effect on the sample previously exposed to pH 8.3 water. For the sample initially at pH 7.7, the maximum E_{corr} increased about 66 mV when raised to pH 9.0, whereas the sample initially at pH 8.3 sample increased about 669 mV when raised to pH 10.0. We believe that final pH is the controlling factor. At present, we do not know whether the pH of initial exposure plays a role, although it could.

Orthophosphate and pH had profound impacts on the visual characteristics of copper corrosion and scale formation (Figure 11.7). Prior to increasing the pH at 7.7 or 8.3, the samples with phosphate developed a blue-green scale, while the same samples exposed to the same water without phosphate had a brown-black scale buildup. When the pH was increased, the appearance of the samples without phosphate remained unchanged; however, the samples with phosphate almost immediately began to turn brownish-black (Figure 11.7). When the sample containing phosphate was raised to pH 10, it began to turn color almost immediately, whereas the sample with phosphate raised to pH 9.0 took several days before a complete change in appearance was observed. From these results, it is obvious that phosphate dramatically alters the scale that is present. In the context of preventing pinhole leaks, it is not certain whether these observations indicate that phosphate loses its effectiveness at higher pH. However, it is clear that at pH 9.0, phosphate has dramatically different effects than in the same water at pH 8.3.

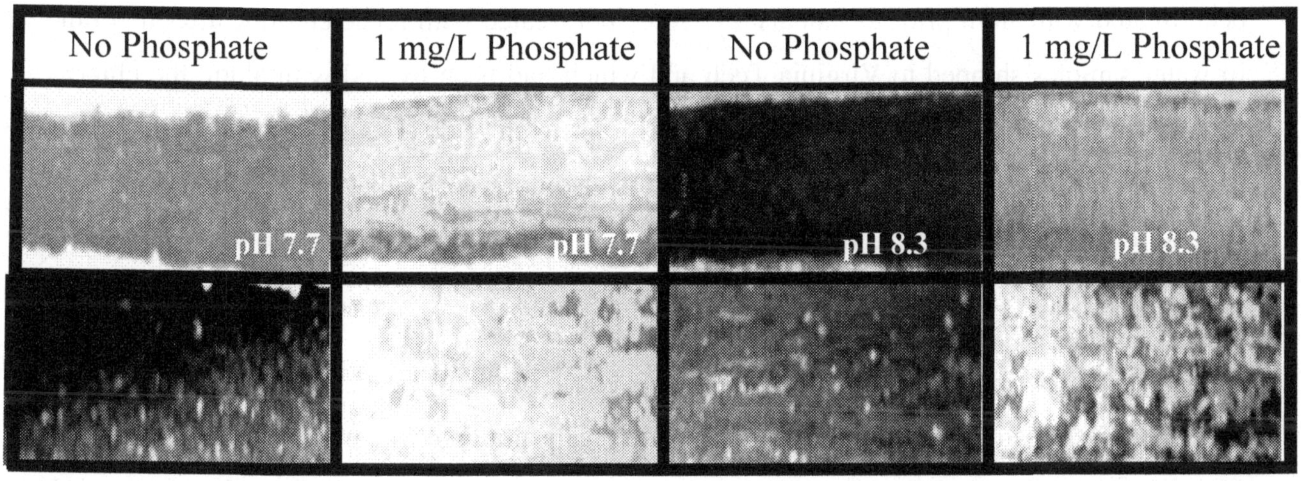

Figure 11.7. Visual comparison of samples before (top row) and after (bottom row) pH increase at end of experiment.

Effects of Orthophosphate and pH on Aspects of Copper Corrosion in Real WSSC Water

A second phase of experiments revealed several key differences between the synthetic and real water. In this test, the copper pipe samples were pre-exposed to the synthetic water containing high levels of chlorine, aluminum and at a pH of 9.2. The goal was to initiate serious pitting of the surface using the principles defined in earlier tests. Thereafter, real WSSC water was circulated through the samples to investigate the effects of orthophosphate on corrosion in this media.

Chlorine Decay Kinetics. Contrary to what was previously observed in synthetic water, there was only a slight difference in chlorine decay rates in WSSC water with phosphate compared to the same water without phosphate (Table 11.2). For instance, the chlorine decay rate was 37 $\mu g/cm^2$ with phosphate versus 42 $\mu g/cm^2$ without phosphate in WSSC water at pH 8.3. This was the first indication that there was something in the real WSSC water that greatly reduced the rate of reaction between free chorine and the copper. While the comparison is not perfect due to small differences in the experimental setup in the synthetic water versus real water tests, between 17-55% lower chlorine decay rates were observed in real WSSC water compared to the synthetic water test at the same pH (Table 11.3). This is even more surprising given that WSSC water likely contains organic matter and other substances that would have a chlorine demand compared to the synthetic water.

Unbeknownst to the research team at Virginia Tech, the real water used in testing had been treated with alum prior to January 21, 2003, but treated with ferric chloride after that time. In water samples shipped to Virginia Tech and which had been treated with alum, the chlorine residual varied more than in the water treated with ferric chloride (Figure 11.8), even though we used the exact same chlorine dosing scheme. It is also noteworthy that pH no longer seemed to have an effect on the chlorine decay rate because all samples seemed to follow the same trend (Figure 11.8).

Corrosion Potential. All copper samples in this phase of the work were previously exposed to the same pit initiation step using synthetic water with aluminum, 5 mg/L chlorine at pH 9.2. Therafter, they were switched to WSSC water at a given target pH and level of phosphate. In terms of E_{corr}, immediately after the switch was made to WSSC water, E_{corr} dropped by approximately 200 mV within the few minutes before we could take or first measurement (Figure 11.9). This could be due to the fact that aluminum was no longer dosed to the WSSC water whereas it was dosed to the synthetic water, the target free chlorine level was lowered from 5 down to 4 mg/L, or certain substances in the WSSC water were responsible for reducing the E_{corr} directly. For reasons that are discussed later, the latter prospect is deemed most likely.

Table 11.2. Chlorine Consumption of Synthetic vs. WSSC Water.

	Consumption (ug/cm^2)		
	Synthetic Water	*WSSC water*	% Reduction
pH 8.3 with P	54.3	37.2	31%
pH 8.3	82.2	41.9	49%
pH 9.1 with P	48.1	39.7	17%
pH 9.1	91.5	41.4	55%

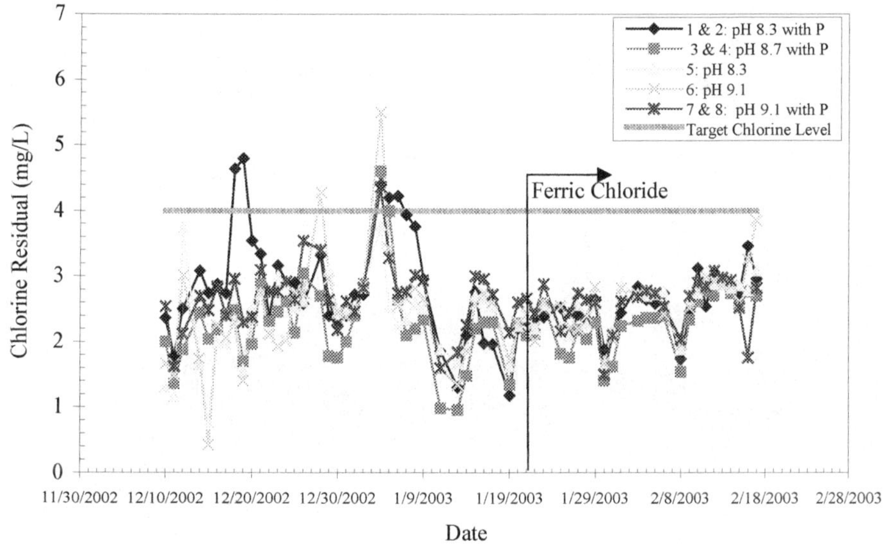

Figure 11.8. Chlorine Residuals Measured Daily.

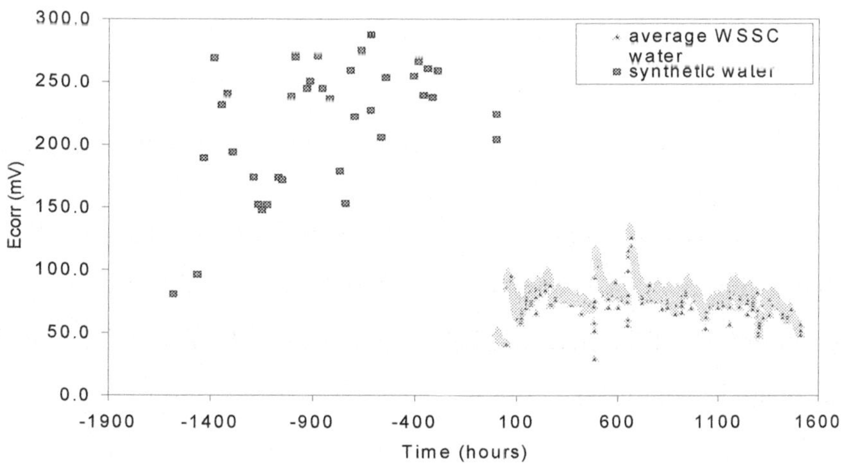

Figure 11.9. Average Ecorr Readings for Initiation Stage with Synthetic Water and Propagation Stage with WSSC Water.

A given sample from WSSC was recirculated throughout the samples for a given period of time (a week or two), before it was replaced with a fresh WSSC water sample. "Spikes" in E_{corr} for a given condition occurred whenever the water was changed and "new" WSSC water was recirculated through the copper samples (Figure 11.10). It was noticed that these temporary spikes in E_{corr} did not occur as regularly after about 1000 hours. Upon reflection, we suspect that this could be due to the change in coagulant used for the WSSC water.

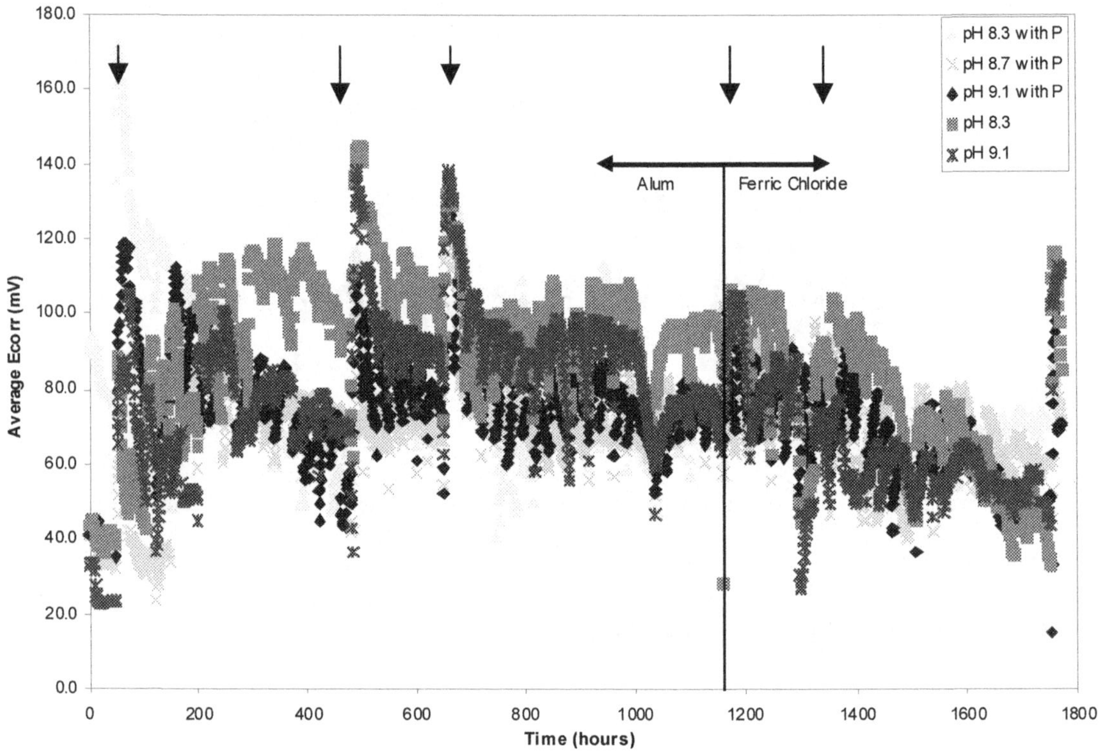

Figure 11.10. Average E_{corr} for samples in WSSC water. Arrows indicate water changes.

Due to the relatively narrow range in E_{corr} values and the complexity of the results, the data was examined more closely in an attempt to extract additional useful information and trends related to phosphate inhibition. If the percent difference between the E_{corr} values of the samples with and without phosphate is calculated as follows:

$$\Delta E_{corr} = (E_{corr} \text{ with phosphate}) - (E_{corr} \text{ without phosphate}) \qquad \text{Equation 11.3}$$

Assuming that a higher E_{corr} is an indication of higher pitting tendencies, if the above difference in E_{corr} is negative, the effect of phosphate is beneficial (Figures 11.11 and 11.12). For both pH 8.3 and 9.1, the presence of P seemed have an initial detrimental effect on the corrosion potential (ie- increased pitting tendencies). Between 300 and 1200 hours, the samples dosed with phosphate had a relative positive effect, or decreased pitting tendencies when to the samples without P. The benefit was more obvious at pH 8.3 than at 9.0, as might be expected based on results in the synthetic water.

After about 1200 hours, there was no evident advantage to using phosphate in WSSC water in the context of Ecorr at pH 8.3, although some benefits were noted in one of the two samples at pH 9.1. These results may be due to a change in water composition. After 1200 hours, we started using the water shipped from WSSC that had been treated with ferric chloride instead of alum as a coagulant. Implications of this are discussed in more detail in the following section.

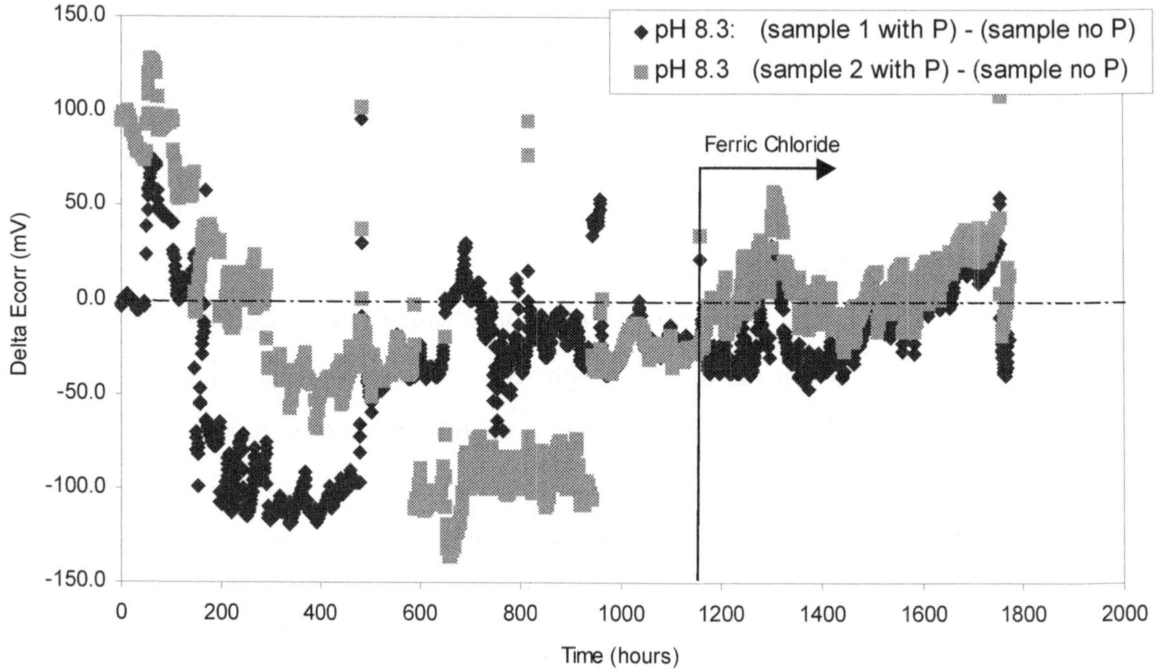

Figure 11.11. Delta E_{corr} Between Samples with and without Phosphate for pH 8.3.

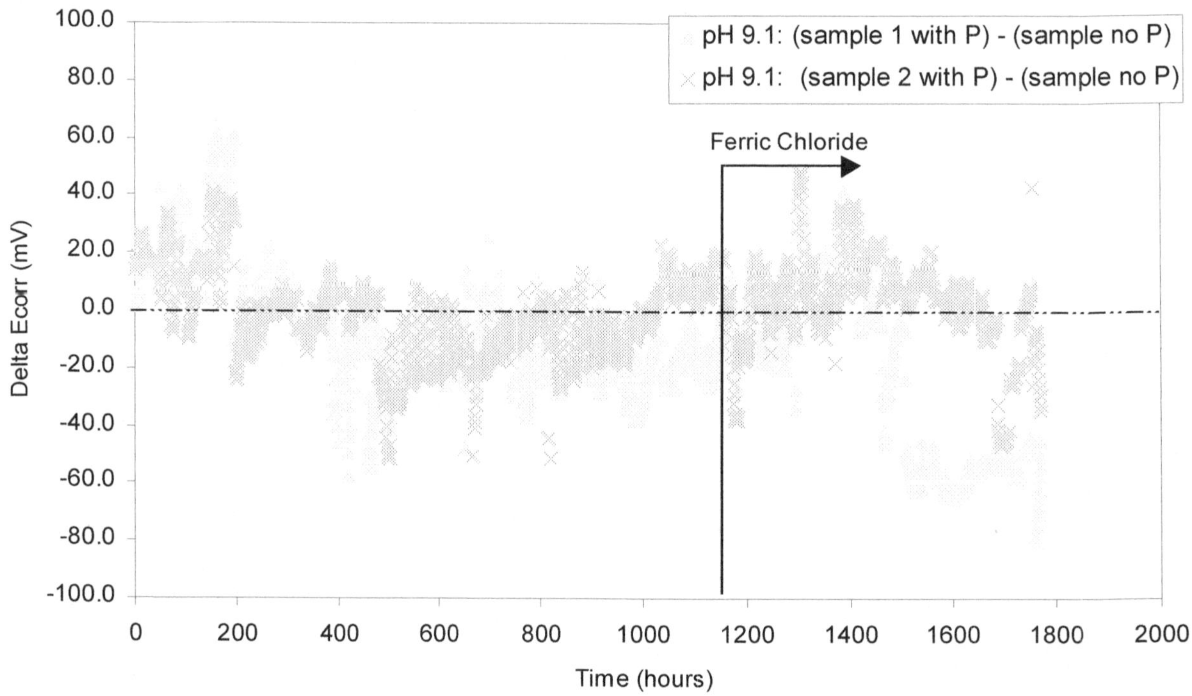

Figure 11.12. Delta E_{corr} Between Samples with and without Phosphate for pH 9.1.

Overall Comparison of Synthetic and Real WSSC Water

Synthesis of data from the two experiments illustrates two dramatic trends. First, in relation to factors controlling pitting tendencies, the WSSC water had remarkably lower pitting propensity than the synthesized water. Because of this, impacts of phosphate were more difficult to distinguish in these short-term tests in real WSSC water. It was also suspected that a change in the pitting propensity of WSSC water occurred in the middle of the study. After the study was completed, it was determined the coagulant had been changed, and this was one possible explanation.

It is useful to compare the composition of the synthetic water to that typical for WSSC water treated by alum and WSSC water treated by ferric chloride (Table 11.3). Alkalinity, Ca^{+2}, sulfate levels are very similar in these waters as was desired based on the design of the experiment. However, organic matter and silica and nitrate were not added to the synthesized water. It is possible that the absence of these trace constituents is mostly responsible for the major difference in corrosivity, especially since it is believed that NOM inhibits copper pitting tendencies (Edwards et al., 1994a).

It is also interesting to note that chloride was 1.7 times higher after the water was treated by ferric chloride. This could also have benefits to pitting, since higher chloride is noted to decrease E_{corr} of copper samples with time, and also reduce pitting tendencies (Edwards et al., 1994b). This is also a potential explanation for decreases in E_{corr} with time during synthetic water and real water experiments. That is, due to reaction 11.1, Cl^- concentration was building up in samples with time as free chlorine was replaced. Based on our measurements, increases in chloride concentration up to 1.5 times were occurring in the synthetic water tests, and not much else was changing in the water with time. Thus, the data support the idea that higher Cl^- in the water after ferric chloride dosing was having effects on E_{corr}.

Another possible contributing factor is that water treated with alum may have had more residual aluminum in it than water treated by ferric chloride. This was not determined during the test, since we did not know in advance that the coagulant would be changed. If aluminum was higher in alum treated water, this could also be a factor, since WSSC water with added aluminum has previously exhibited greater pitting tendencies in the presence of chlorine when compared to the same water without extra aluminum (Edwards, 2002). We do not have any data as to effects of ferric iron on rates of chlorine decay in the presence of WSSC water or in synthetic water

Finally, the relative merits of ferric chloride versus alum have been manifested for copper corrosion by-product release in previous work (Rerhing et al., 1996), although this was attributed to the difference in chloride versus sulfate in the final water.

Regardless of the exact causes, it may explain the reduced benefits from phosphate dosing to real WSSC water. While phosphate was nearly always beneficial, it did require time before effects were observed, and the data had to be examined very closely to find significant trends.

The relative pitting propensity of WSSC versus the synthesized water is also obvious based on simple visual observations of the samples. In the synthesized water, tens of protuberances (possible tubercles) were present in an inch long section of copper coupling (Figure 11.13). We believe that all these tubercles were formed in the synthetic water, and not in the WSSC water exposure, and this is the reason there is not an obvious visual difference between the condition with and without phosphate. In forensic analysis of failures from homes of WSSC customers, even one tubercle per inch of pipe would be deemed extraordinarily high. In summary, while it is believed that the trends noted in synthesized water are directly relevant to the problem at WSSC, trace constituents such as NOM are probably playing as important a role in controlling pitting as does the added orthophosphate. In some ways, dosing of orthophosphate is probably only replacing a portion of this "natural" inhibiting capability, and maybe only for waters at pH 8.3. Indeed, it is quite possible that use of ferric chloride coagulant instead of alum or polyaluminum chloride might have more benefits to pitting frequency than dosing of orthophosphate in this water, but this must be considered speculative at this time. Under the circumstances, any benefits from orthophosphate would be highly desirable, and its dosing is therefore recommended.

Table 11.3. Chemical Composition Comparison of Synthetic and WSSC Water.

	units	*Synthetic Water*	*WSSC water (Alum)*	*WSSC water (Ferric Chloride)*
Alkalinity as $CaCO_3$	mg/L	34	33-38	30-36
Cl^-	mg/L	20	20-25	36-41
Ca^{+2}	mg/L	17	17-18	17-20
NO_3 as N	mg/L	0	2.5	2.5
Silicon	mg/L	0	2.9-3.2	3-3.6
Sulfate	mg/L	14	7.7-13	13.4
TOC	mg/L	0	1.7-2	1.6-2.2

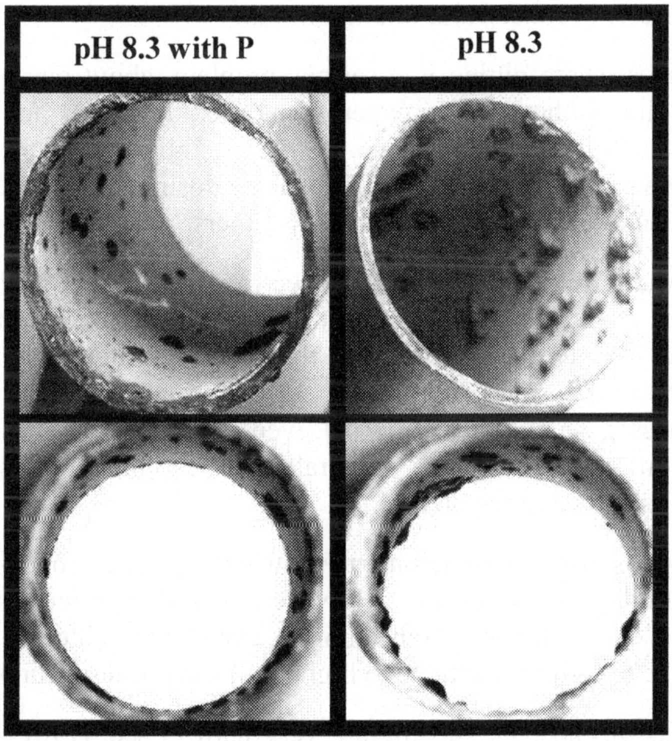

Figure 11.13. Visual Comparison of Tubercle Formation.

The following conclusions are drawn from experimental results.

- There is a natural inhibitor to pitting corrosion in WSSC water that is not present in the synthetic water. Silica and NOM seem to be the most likely candidates.
- When we put fresh WSSC water into the rig, E_{corr} spiked up, before slowly coming down during as the test progressed. We interpret this to indicate that in our tests with WSSC water, either something was building up during the test that tended to decrease pitting intensity (as measured by E_{corr}), or something that was promoting pitting was being removed from the water. It is known that chloride was building up during the test. It is unknown what was happening to aluminum levels in the water during the test. These seem to be two likely possible explanations for the phenomena.
- The higher chloride concentration in the water after ferric chloride was dosed at the treatment plant may have reduced the pitting propensity of the water. This could be due to higher chloride in the water, or reduced aluminum, or both.
- In real WSSC pipes where pitting is a serious problem, about one pit is observed every 3 inches or so. In the synthetic water experiments (i.e., Table 8), tens of possible tubercles were formed per inch in a period of one month. It follows that the relatively low pitting density in WSSC water makes benefits of phosphate dosing harder to detect, even though the rate of pitting is unacceptably high to WSSC customers.
- In the synthetic water at pH 7.7, we did not see much evidence of pitting, as was expected based on previous results. Phosphates definitely helped reduce the pitting intensity in the synthetic water at the pH of 8.3 compared to synthetic water without phosphate. This is deemed very important, since distributed WSSC water is often near pH 8.3 or less. Phosphates had lesser benefits at higher pH even in synthetic water, but they were clearly having impacts on scale formation. Overall, even at higher pHs, some benefits from orthophosphate dosing are anticipated.
- In WSSC water, phosphates were helping to reduce the pitting propensity of the real water at pH 8.3 after a period of days exposure. However, even though the result must be deemed of low confidence, if the water had high chloride from ferric chloride treatment, benefits of phosphate might be less significant due to higher chloride or less aluminum in the water.
- At pH 9.1 benefits of phosphate dosing are very hard to see. However, the trend is toward slight benefits with time.
- As shown in the data, water treated by alum and ferric chloride exhibit major differences in non-uniform copper corrosion behavior.

REFERENCES

AWWARF & DVGW-TZW. Internal Corrosion of Water Distribution Systems, 2nd edition. AWWARF, Denver, CO (1996).

Broomfield, J.P. Carbonation and Its Effects in Reinforced Concrete. *Materials Performance.* January (2000)

Edwards, M. Understanding the Problem of Pinhole Leaks in Homes of WSSC Consumers. Various Internal Reports for the Washington Suburban Sanitary Commision (2002).

Edwards, M., J. Rehring and T. Meyer. Inorganic Anions and Copper Pitting. ***Corrosion,*** V. 50, No. 5, 366-372 (1994b).

Edwards, M., J.F. Ferguson and S. Reiber. The Pitting Corrosion of Copper. ***Journal American Water Works Association***. V. 86, No. 7, 74-90 (1994a).

Hooton, R. D. Prediction of Chloride Penetration in Concrete. Publication No. FHWARD-00-142. Oct. 2001

Kireta, A. Personal Communication February 21, 2003.

Otsuki, N; Nagataki, S; Nakashita, K. Evaluation of AgNO3 Solution Spray Method for Measurement of Chloride Penetration into Hardened Cementitious Matrix Materials. *ACI Materials Journal.* **1992**, Nov-Dec., 587.

Rehring, J. P, and M. Edwards. Copper Corrosion in Potable Water Systems: Impacts of Natural Organic Matter and Water Treatment Processes. ***Corrosion***. V. 52, No. 4 307-317 (1996).

Reiber, S.H., *Copper Plumbing Surfaces: An Electrochemical Study.* Journal of the American Water Works Association, 1989. 81(7): p. 114-122.

Rushing, J.C., and M. Edwards. Role of Chlorine and Aluminum in Pitting Corrosion of Copper Pipes. Proceedings of the 2002 Water Quality Technology Conference in Seattle Washington. November, 2002b.

Rushing, Jason C., Virginia Tech M.S. Thesis (2002a).

APPENDIX

Table 1A. Aggregate Specifications

Coarse aggregate, crushed siliceous gravel, maximum nominal size ½ inch.

Percent Passing

Size	Crushed Gravel	ASTM C 33 #7
3/4 in.	100	100
1/2 in.	100	90-100
3/8 in	50	40-70
#4	8	0-15
#8	2	0-5

Specific Gravities:
Bulk Dry = 2.59
Bulk S.S.D. = 2.62
Apparent = 2.65

Absorption % = 0.81
Dry Rodded
Unit Weight = 100 lbs/cf

Fine aggregate, natural sand.

Percent Passing

Size	**Natural Sand**	**ASTM C 33**
3/8 in.	100	100
#4	99	95-100
#8	90	80-100
#16	78	50-85
#30	46	25-60
#50	17	10-30
#100	2	2-10
#200	0.5	

Specific Gravities:
Bulk Dry = 2.55
Bulk S.S.D. = 2.57
Apparent = 2.60

Absorption % = 0.75
Fineness Modulus = 2.68

Table 2A. Portland Cement

Cement Type I/II

Chemical Analysis

Analyte	Weight Percent	ASTM C 150 **Type II**
Si O_2	21.25	20.0 min
Al_2O_3	4.49	6.0 max
Fe_2O_3	3.04	6.0 max
CaO	63.51	-
MgO	2.48	6.0 max
SO_3	2.47	3.0 max
Na_2O	0.17	-
K_2O	0.82	-
TiO_2	0.21	-
P_2O_5	0.11	-
Mn_2O_3	0.06	-
SrO	0.14	-
L.O.I. (950° C)	1.06	3.0 max
Alkalies as Na_2O	0.72	0.60 max

Table 3A. Concrete Mixture Proportions

ACI 211.1-85 Standard Practice for Selecting Proportions for Normal, Heavyweight, and Mass Concrete

Batch Weights

	Batch Weights, lbs/cy		
	Oven-dry	SSD	
Ingredient	lbs	cf	lbs
Cement, Type I/II	**592**	**3.01**	**592**
Water	**249**	**3.99**	**249**
Coarse Aggregate	**1512**	**9.35**	**1524**
Fine Aggregate	**1310**	**8.23**	**1318**
Air Content	**-**	**2.42**	**-**
		27.00	**3683**

Slump = **3 to 4 in.**
Air Content = **6 to 9%**
W/C = **0.42**
Unit Weight = **136.4 lbs/cf**
Fine/Total Aggregate solid volume ratio = **0.47**

Table 4A. Concrete Mixture Batch Quantities, Fresh and Hardened Properties

	S.S.D Weights, lbs/cg		
Ingredient	Batch 1 lbs	Batch 2 lbs	Specification FCWA
Cement	**592**	**592**	**592 min**
Water	**249**	**249**	**-**
Coarse Aggregate	**1524**	**1524**	**-**
Fine Aggregate	**1318**	**1318**	**-**
Daravair	**7.4 oz.**	**4.7 oz.**	**-**
Daracem 19	**148 oz.**	**106 oz.**	**-**
Concrete Temp. ^oF	**66**	**65**	**45 min**
Slump, in.	**9**	**5**	**8 max**
Air Content, %	**12.0**	**8.0**	**5-7**
Unit Weight, lbs/cf	**131.2**	**139.2**	**-**
Relative Yield, %	**104%**	**0.98**	**-**
W/C	**0.42**	**0.42**	**0.42 max**
Fine/Total Aggregate	**0.47**	**0.47**	**40-55**

Compressive Strength:			
7 day, 3000 + 1200	5330	4810	4200 psi
28 day, 4000 + 1200			5200 psi

IWA Publishing's authorised EU representative for General Product Safety Regulations is Diane D'Arras, 15 rue Duret, 75116 Paris, France, e-mail: safety@iwap.co.uk.

Printed and bound by CPI Group (UK) Ltd, Croydon, CR0 4YY
17/09/2026
02218810-0020